RECOGNIZING CATASTROPHIC
INCIDENT WARNING SIGNS
IN THE PROCESS INDUSTRIES

This book is one in a series of process safety guideline and concept books published by the Center for Chemical Process Safety (CCPS). Please go to *www.wiley.com/go/ccps* for a full list of titles in this series.

RECOGNIZING CATASTROPIC INCIDENT WARNING SIGNS
IN THE PROCESS INDUSTRIES

CENTER FOR CHEMICAL PROCESS SAFETY
of the
AMERICAN INSTITUTE OF CHEMICAL ENGINEERS

American Institute of Chemical Engineers
3 Park Avenue
New York, NY 10016

A JOHN WILEY & SONS, INC., PUBLICATION

Published by John Wiley & Sons, Inc., Hoboken, New Jersey.
Published simultaneously in Canada.

For general information on our other products and services please contact our Customer Care Department within the United States at (800) 762-2974, outside the United States at (317) 572-3993 or fax (317) 572-4002.

Wiley also publishes its books in a variety of electronic formats. Some content that appears in print, however, may not be available in electronic formats. For more information about Wiley products, visit our web site at www.wiley.com.

Library of Congress Cataloging-in-Publication Data:

Recognizing catastropic incident warning signs in the process industries / Center for Chemical Process Safety.
      p. cm.
   Includes index.
   ISBN 978-0-470-76774-0
   1. Chemical engineering—Safety measures. 2. Chemicals—Accidents—Prevention. 3. Warnings. I. American Institute of Chemical Engineers. Center for Chemical Process Safety.
   TP150.S24R44 2012
   660'.2804—dc23   2011033614

Printed in the United States of America.

10 9 8 7 6 5 4 3 2

It is sincerely hoped that the information presented in this document will lead to an even more impressive safety record for the entire industry; however, neither the American Institute of Chemical Engineers, its consultants, CCPS Technical Steering Committee and Subcommittee members, their employers, their employers' officers and directors, nor AntiEntropics, Incorporated (AEI) and its employees warrant or represent, expressly or by implication, the correctness or accuracy of the content of the information presented in this document. As between (1) American Institute of Chemical Engineers, its consultants, CCPS Technical Steering Committee and Subcommittee members, their employers, their employers' officers and directors, and AntiEntropics, Inc., and its employees, and (2) the user of this document, the user accepts any legal liability or responsibility whatsoever for the consequence of its use or misuse.

CONTENTS

6   PROCEDURES   79

LIST OF TABLES

LIST OF FIGURES

FILES ON THE WEB ACCOMPANYING THIS BOOK

Access the incident warning sign self-assessment tool and a list of the warning signs using the Microsoft Explorer Web browser at

http://www.aiche.org/ccps/publications/IWSMaterial.aspx

Password: IWS2010

ACKNOWLEDGMENTS

The American Institute of Chemical Engineers (AIChE) wishes to thank the Center for Chemical Process Safety (CCPS) and those involved in its operation, including its many sponsors whose funding made this project possible, and the members of the Technical Steering Committee, who conceived of and supported this concept book project. The members of the CCPS Incident Warning Signs Subcommittee who worked with AntiEntropics, Inc. to write this text deserve special recognition for their dedicated efforts, technical contributions, and overall enthusiasm for creating a useful addition to the concept book series. CCPS also wishes to thank the subcommittee members' respective companies for supporting their involvement in this project.

The co-chairpersons of the Incident Warning Signs Subcommittee were Joyce Becker of BP and Ronald Rhodes of Total Petrochemicals USA, Inc. The CCPS project manager was Brian Kelly. The members of the CCPS subcommittee were:

- Steve Arendt   *ABS Consulting*
- Todd Aukerman   *LANXESS Corporation*
- Larry Bowler   *SABIC Americas, Inc.*
- Michael Boyd   *Husky Energy*
- Owen Chappel   *BP*
- Robert Fischer   *Total Petrochemicals USA, Inc.*
- Kevin He   *Dow Corning Corporation*
- John Herber   *CCPS Emeritus*
- James Klein   *DuPont*
- David Lewis   *Occidental Chemical Corporation*
- Kevin MacDougall   *Husky Energy*
- Doug Morrison   *Nexen Inc.*
- John Murphy   *CCPS Emeritus*
- Charles Pacella   *Baker Engineering and Risk Consultants, Inc.*
- Fred Simmons   *Savannah River Nuclear Solutions, LLC*
- Jim Slaugh   *LyondellBasell Industries*

Robert Walter, president of AntiEntropics, Inc., was the lead writer for this book. Sandra A. Baker of AntiEntropics, Inc. was the editor. Richard Foottit, Kerry Fritz, and Cliff Van Goethem of AntiEntropics, Inc. provided internal team review. In addition, AntiEntropics would like to recognize the entire CCPS team for their writing contributions throughout the book.

CCPS also gratefully acknowledges the comments submitted by the following peer reviewers:

- John Alderman   *Aon Consulting*
- Martyn Fear   *Husky Energy (Offshore Operations)*
- Andy Hart   *Nova Chemical*

- Dennis Hendershot   *CCPS Emeritus*
- Gregg Kiihne   *BASF Corporation*
- R. Craig Matthiessen   *US Environmental Protection Agency*
- Louisa A. Nara   *CCPS*
- Robert Ormsby   *CCPS Emeritus*
- Stephen Selk   *US Department of Homeland Security*
- Kenneth Wengert   *Kraft Foods Global, Inc.*
- David Worthington   *Amerada Hess*
- David Wulf   *ConocoPhillips*

Their insights, comments, and suggestions helped ensure a balanced perspective. Although the peer reviewers have provided many constructive comments and suggestions, they were not asked to endorse this book and were not shown the final draft before its release.

The photograph used for the cover and as Figure 2-8 is reproduced with permission from the Associated Press.

FOREWORD

The oil spill from the Macondo well in the Gulf of Mexico in 2010 was a classic example of the thesis of this book. In the lead up to the accident there were numerous warning signs, some subtle, but some utterly unambiguous. Most strikingly, in the hours immediately before the blowout, the well was giving unmistakable signs that it had not been properly sealed. These were either missed, because people had stopped paying attention in the rush to get the job finished, or dismissed, because ways were found to normalize them. Hours earlier there were a number of other anomalies. The meaning of these anomalies was less clear at the time, but in retrospect were probably signs that something was wrong. These were passed over without adequate thought as to their implications.

Warning signs can sometimes be used as the basis for safety indicators. For instance, a well instability event (often called a *kick*) is a warning of danger, and the number of such events could possibly be used as the basis for a safety indicator. However, well integrity was not considered a safety matter in the Gulf of Mexico and so no such safety indicators had been developed. Safety with respect to major hazards depends on developing such indicators and incorporating them into management systems.

This book catalogs a large range of warning signs that are worthy of attention. It is an extremely useful source for people seeking to develop key performance indicators (KPIs) for how well major hazard safety is being managed. The book also discusses some of the reasons why these signs are passed over so often and what needs to be done to ensure that we pay them proper attention. As such, it is a valuable addition to safety literature in hazardous industries.

Andrew Hopkins
Emeritus Professor of Sociology, Australian National University
March 2011

PREFACE

The American Institute of Chemical Engineers (AIChE) has been closely involved with process safety and loss control issues in the chemical and allied industries for more than four decades. Through its strong ties with process designers, constructors, operators, safety professionals, and members of academia, AIChE has enhanced communications and fostered continuous improvement of the industry's high safety standards. AIChE publications and symposia have become information resources for those devoted to process safety and environmental protection.

AIChE created the Center for Chemical Process Safety (CCPS) in 1985 after the chemical disasters in Mexico City, Mexico, and Bhopal, India. The CCPS charter is to develop and disseminate technical information for use in the prevention of major chemical catastrophic incidents. The center is supported by more than 135 chemical process industry (CPI) sponsors who provide the necessary funding and professional guidance to its technical committees. The major product of CCPS activities has been a series of guidelines and concept books to assist those implementing various elements of a process safety and risk management system. This book is part of that series.

The CCPS Technical Steering Committee initiated the creation of the concept books and guidelines to assist facilities in meeting these challenges. This book contains approaches for continually improving a process safety management system and developing the culture necessary to implement it. The Web files accompanying this book contain resource materials and support information.

Process safety programs to protect the lives of workers and the public deserve the
same level of attention, investment, and scrutiny as companies
now dedicate to maintaining their financial controls.
Carolyn Merritt (1947 – 2008)
Former United States Chemical Safety Board Chairman

1
INTRODUCTION

Warning signs are indicators that something is wrong or about to go wrong. When we recognize and act on these indicators, a loss may be prevented. Of course, this will only happen when we know what to look for and are willing to take the initiative to do something about it. A review of significant incidents in the process industries suggests that most if not all incidents were preceded by warning signs. Some of these signs were clearly visible but not acted upon because their significance was not understood. Other warning signs were less obvious, but observant personnel may have detected them.

This book is about warning signs that have preceded or contributed to past incidents. There is one common characteristic shared by the incident warning signs presented here:

The organization does not perceive or recognize them.

An incident warning sign is a subtle indicator of a problem that could lead to an incident. Some minor incidents might have the potential to escalate to a catastrophic level. There are warning signs of a physical or tangible nature, and there are warning signs related to the management practices of an organization. Some warning signs may be problems in themselves while others may be symptoms of potential problems or incidents. Every sign provides a clue that may be an early warning of catastrophe. These clues give us an opportunity to do things differently to reduce the risk of a catastrophe.

It is common for an organization that experiences a major catastrophic incident to respond initially with shock and surprise. Somehow, facility workers and their supervisors may have developed the impression that catastrophic incidents only happen elsewhere and are the result of gross misconduct or total system breakdown. As the organization or local facility may have sustained itself without a catastrophic incident for many years, managers, technical staff, and facility workers often hold the sentiment...*We must be doing things right.* In many cases, this is a misguided perception. Consider taking a closer look at possible warning signs at your facility to avoid this experience.

The following quote from Professor Andrew Hopkins, an author on safety and professor of sociology at Australian National University, emphasizes the importance of addressing these indicators.

> *Prior to any major accident there are always warning signs which, had they been responded to, would have averted the incident. But they weren't. They were ignored. Very often there is a whole culture of denial operating to suppress these warning signs.*

The early warning signs we discuss either are unknown to the organization due to their lack of visibility or because the associated risk was unrecognized, downplayed, or ignored.

There are many types of warning signs, including the following:

- Early indicators of failure that provide opportunities to take appropriate action. Process equipment that is not functioning properly may be prone to failure. Organizations sometimes ignore these on the basis that they will address the issues later (or if the problem escalates).
- Suggestions that a major incident may be imminent. An example might include a piece of process equipment reaching its end-of-cycle or retirement limit.
- Indicators of other issues that are less obvious and require detailed analysis. For this reason, a practical follow-up option is to conduct an audit to help ensure that programs and systems are managed effectively.

- Seemingly insignificant issues that, when combined with other warning signs, suggest a breakdown of management systems.

- Actual incidents with measurable consequences. If we ignore these, they can increase in frequency or magnitude and contribute to a catastrophic incident.

Some of the warning signs discussed in this book are commonplace across the process industries. They may exist within your own operation or facility. Other significant warning signs may have been missed in our analysis and you may need to augment the list to make it more suitable to your organization's needs. It would be rare to find a facility that does not have a warning sign present. When any warning sign becomes widespread or becomes the norm within an organization or when multiple warning signs are present, there is reason for concern. However, some warning signs might be inherent in your type of operation, and total elimination may be impossible or impractical.

Although some warning sign discussions do offer suggested approaches, the goal of this book is not to explain *how to do it* or *how to fix the problem*. We hope this book is a trigger for initiating self-assessment and follow-up action. It should help to alert responsible leaders that trouble may be on the way and, unless these warning signs are recognized and addressed, the probability of experiencing a major incident could increase.

1.1 PROCESS SAFETY MANAGEMENT

Process safety management (PSM) is a systematic framework of activities used to manage the integrity of hazardous operations and processes. Specifically, it deals with the prevention of *loss of containment* incidents that are commonly associated with the process industries. When hazardous materials are released outside their primary containment, especially to the atmosphere, the effects can be widespread and difficult to predict. Prevention is of paramount importance. Initiating proactive and systematic response to incident warning signs is a powerful tool to augment the standard elements of process safety. Determine your organization's process safety philosophy. Ask the following question of your organization.

> ***Do our actions related to process safety reflect that process safety is an integral part of our organizational culture?***

Organizations that understand the importance of process safety will want to reflect that awareness in their actions. To be most effective, embrace rigorous process safety as a value throughout the organization and show it in the organization's actions. Priorities change with the working environment or business climate. Values do not. They are something that a group internalizes and characterizes in its behavior to help define a framework for its actions.

1.1.1 Identifying process safety management system deficiencies

Less serious incidents precede most catastrophic incidents. Many of these less-than-catastrophic incidents can reveal early warning signs that could have prevented the catastrophic incident from occurring if leaders had recognized and acted upon them. Management system deficiencies are a root cause in all effective incident investigation methodologies. Our goal should be to identify common management system deficiencies and their associated early warning signs. We use case studies to illustrate these warning signs, and we hope to help you make a connection from these examples to your facilities. The case studies chosen are landmark events supported by thorough investigations. These catastrophes influenced how companies manage process safety activities worldwide.

Industry generally understands the central concept of designing and implementing management systems to maintain and improve performance. Process safety management embodies the following principles, although you may not recognize these as elements in your company's program:

- Know your processes—both physically and in terms of information about them.
- Communicate process hazards effectively to all affected employees and contractors.

- Analyze the processes as necessary to understand them and their associated hazards.
- Operate the processes within their safe operating envelopes.
- Maintain the equipment properly in accordance with a documented asset integrity program.
- Manage and communicate all process changes over time.
- Train your employees and contractors on the processes and the associated hazards.
- Evaluate the operation regularly to see how you are doing.
- Get out into your facility so you can observe the visible warning signs and encourage site leadership to do the same.
- Develop the competencies and resources required by your business processes and process safety management system.
- Modify your business processes and management system as needed to meet changing requirements.
- Communicate the status of the facility's activities and findings from any periodic reviews to the site senior management and others as appropriate.
- Develop action plans to address and close all findings.

When all workers see that the organization values and supports these types of systems, it encourages the employee's individual commitment to safety and environmental excellence. The Center for Chemical Process Safety (CCPS) publication *The Business Case for Process Safety* demonstrates that a rigorous process safety framework also improves the bottom line.

Many of these management systems may already be in place in a high-hazard facility or operation. It is wise to consider extending these to all parts of an operation, regardless of the process hazards.

1.2   NORMALIZATION OF DEVIANCE

Normalization of deviance is a long-term phenomenon in which individuals or work teams gradually accept a lower standard of performance until the lower standard becomes the norm. It is typically the result of conditions slowly changing and eroding over time.

Normalization of deviance often begins in the form of a shortcut or poorly documented temporary change in a standard work practice or procedure. If there are no apparent negative consequences, the new practice becomes accepted and displaces the original practice. Over time, this process repeats. When the changes are small and seemingly insignificant, they are easy to miss. One subset of normalized deviance is organizational creep. This is a common term for this type of incremental change away from set work practices in an operating organization. In the midst of such change, it is sometimes difficult to recognize or appreciate the value of the original practice or procedure. Even when discussing this situation with local management, the news may be met with opposition in the form of comments like—*it works just fine*—or—*it has not caused us any problems yet.*

Normalization of deviance represents a serious breach in the management of change system. It is also a major contributor to many serious incidents.

1.3 A STRATEGY FOR RESPONSE

The response upon recognition of incident warning signs at your facility should not be a knee-jerk reaction but rather a disciplined approach to collecting all the facts and moving forward in a planned and organized manner. Figure 1-1 illustrates an approach to responding to an incident warning sign.

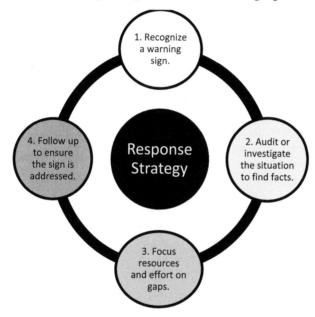

Figure 1-1   A Strategy for Response

There may be a tendency in some organizations to deny the importance or significance of certain warning signs. This may be the case when these warning signs have existed for a considerable period without consequence. Some facilities may not consider a warning sign important if the entire staff has a high level of seniority and experience, and they have become complacent about plant operations. This indicates a false sense of security. Both new and senior workers may need training on the correct methods for doing things so that they develop a heightened sense of awareness regarding the importance of warning signs.

Professor Andrew Hopkins further states, "Organizational culture is one of the key factors why companies fail to recognize the warning signs prior to workplace accidents."

An analysis of actual incidents may suggest that certain early warning signs correlate with specific types or categories of incidents. This assumption may not be valid. Managers need to take time to examine the local trends. They should stay close to day-to-day operations and not be lulled into complacency by short-term successes.

Finding multiple warning signs within your organization provides a more compelling reason to be concerned and to take action. Carefully screen the operation to determine system gaps or weaknesses. If these exist, are they universal or localized in one part of the operation? For example, procedural weaknesses can result in poor training. Likewise, inadequate training can result in inadequate written procedures. Analyze each detected sign carefully to ensure that you are fixing the right problem.

- Some of the warning signs may appear similar in context. While it is tempting to combine these, each might have a unique explanation or cause.
- Incident warning signs are unlikely to exist in isolation. Where a significant warning sign exists, chances are that other warning signs are present as well.
- It may be possible to rationalize a system weakness, but it is difficult to explain why so many things have gone wrong. This is why it is advisable to investigate further before initiating action.

Ultimately, all employees and contractors at a facility play a role in ensuring that the organization follows its programs rigorously and addresses all gaps. Addressing incident warning signs is a practice that should accompany other established initiatives within a company's process safety framework. Recognizing and understanding these warning signs will better prepare you and your co-workers to take on that task. Success in achieving this goal depends on the combination of well-understood process safety guidelines and operational discipline.

The questions in Table 1-1 support the recognition and follow-up for these early incident warning signs. Ask these questions whenever a warning sign is detected.

TABLE 1-1 Key Questions for Warning Sign Incident Investigations

Has this warning sign been observed previously?
If so, was it acted upon with satisfactory results?
Is this warning sign evident in all operating areas or particular parts of the operation?
Is there a rational explanation for this warning sign that would suggest that the operation is not really at risk?
Are there other coexisting incident warning signs?
Are the coexisting warning signs related?
If so, how are they related?
Is there a common theme linking the warning signs? (For example, a particular staff person or team, a corporate policy reinforcing the existence of warning signs, or funding cuts.)
Is there any action or activity that might alleviate or eliminate the warning signs noted?

1.4 MAINTAINING ORGANIZATIONAL MEMORY AND A HEALTHY SENSE OF VULNERABILITY

A process safety management system that is well documented and maintained will augment an organization's memory. Up-to-date piping and instrument diagrams (P&IDs), accurate procedures, hazardous area classification drawings, and a thoroughly documented and implemented management of change and pre-startup safety review system are prime examples of how the system creates memories. Maintenance of organizational memory should be reinforced continuously throughout the life cycle of a processing facility.

Maintaining accurate configuration data in real time and regular emergency drills are common practice in the commercial nuclear power industry. This promotes the concept of a healthy sense of vulnerability. Management encourages everyone to maintain the calm sense of awareness that—*the worst-case scenario can happen right now...we need to be ready to follow our response procedures.* Extensive training, strict document control for all critical procedure performance, configuration management, and other good practices keep this in the front of each employee's mind. A healthy sense of vulnerability is an attitude to consider developing for your organization. Do not rely on an organization's history of no catastrophic incidents as a sign of good performance. That history may be a function of good luck.

1.5   RISK BASED PROCESS SAFETY

Since the initial introduction of formal regulations in many countries, process safety appears to have stagnated in some companies. Citing perceptions that—*we experience lower than normal industry risk*—or—*we have a clean record regarding catastrophic incidents*—as possible excuses, some companies have chosen to scale down or compromise the scope of their process safety activities. To address this issue, CCPS has adopted a 20-element risk-based framework for process safety that better addresses the needs of all industry sectors, including those with and without high risks. CCPS member companies developed this framework based on a causal analysis of several major incidents that have occurred since the early 1990s.

Risk based process safety is not a way to circumvent the basic tasks or regulatory elements of process safety as they apply to your facility. Adopting risk based process safety means that your organization must meet the applicable regulations for all of your processes and then look longer, deeper, and in more detail at those processes or parts of processes that present the greatest risk of catastrophic incident. It also involves looking harder at seemingly innocuous utilities such as steam supply, electrical distribution, cooling water, or nitrogen that, in some processes, are critical to bringing a unit back to a safe mode after an excursion. The CCPS risk-based framework helps organizations set up and administer a more effective process safety management system. Through risk based process safety, organizations can focus resources more effectively. This reduces the potential for warning signs to exist. When warning signs do appear, risk based process safety can provide more effective response and correction.

The elements of risk based process safety are shown in Table 1-2.

TABLE 1-2 Risk Based Process Safety's (RBPS) Four Pillars and Twenty Elements

RBPS PILLARS AND ELEMENTS
RBPS PILLAR 1—COMMIT TO PROCESS SAFETY
PROCESS SAFETY CULTURE COMPLIANCE WITH STANDARDS PROCESS SAFETY COMPETENCY WORKFORCE INVOLVEMENT STAKEHOLDER OUTREACH
RBPS PILLAR 2—UNDERSTAND HAZARDS AND RISK
PROCESS KNOWLEDGE MANAGEMENT HAZARD IDENTIFICATION AND RISK ANALYSIS
RBPS PILLAR 3—MANAGE RISK
OPERATING PROCEDURES SAFE WORK PRACTICES ASSET INTEGRITY AND RELIABILITY CONTRACTOR MANAGEMENT TRAINING AND PERFORMANCE ASSURANCE MANAGEMENT OF CHANGE OPERATIONAL READINESS CONDUCT OF OPERATIONS EMERGENCY MANAGEMENT
RBPS PILLAR 4—LEARN FROM EXPERIENCE
INCIDENT INVESTIGATION MEASUREMENT AND METRICS AUDITING MANAGEMENT REVIEW AND CONTINUOUS IMPROVEMENT

A more detailed description of each of the elements can be found in the CCPS guideline series book *Guidelines for Risk Based Process Safety*.

A process safety management system involves a wide range of activities. A performance gap or void in any one of these can increase the likelihood of an incident. When voids exist in several elements, the chance of a major incident is higher. To illustrate this point, Table 1-3 lists several noteworthy incidents that have occurred throughout industry over the past three decades along with the management system deficiencies that contributed to the event. Very few incidents, if any, have resulted from just one cause.

TABLE 1-3 Catastrophic Incident by Element Analysis

Incident	Year	Leadership and Culture	Process Safety Information	Hazard ID & Risk Analysis	Management of Change	Operating Procedures	Training	Incident Investigation	Contractor Management	Emergency Preparedness	Pre-startup Safety Review
Phillips 66 Explosion	1989	X	X	X	X	X	X		X	X	
NASA *Challenger*	1986	X		X	X				X		X
Piper Alpha Platform	1988	X		X	X	X	X		X	X	X
Chernobyl Explosion	1986	X	X	X	X	X	X			X	X
Longford Explosion	1998	X	X	X	X	X	X	X		X	
Jilin, China Explosion	2005	X		X	X	X	X			X	
Bhopal Toxic Gas Release	1984	X	X	X	X	X	X	X		X	
Hickson-Welch Jet Fire	1992	X	X	X	X	X	X		X	X	

Process safety audits measure compliance to process safety management (PSM) goals and standards and pinpoint deficiencies. Audits can be very effective in revealing these deficiencies. However, they are time consuming and require considerable planning. Considering the time and effort required to conduct a formal process safety audit, a facility may be vulnerable during the span between audits. The practical question to answer is—*how does a front line supervisor or manager quickly pinpoint problems or gaps and take appropriate action?* Incident warning signs provide such an opportunity. They are subtle indicators that suggest a need to look further and possibly tighten up or alter process safety practices. Incident warning signs are not to be confused with the components or sub-elements of process safety.

1.6   OUR TARGET AUDIENCE

Our goal is to appeal to a wide range of readers. We especially hope to reach the team leaders, front line supervisors, and production leaders in all process industries. A partial list of these industries is provided below.

- Chemical processing
- Oil and gas production
- Oil refining
- Petrochemical production and processing
- Fossil power generation
- Nuclear power generation
- Pulp and paper
- Pharmaceutical
- Food processing
- Biofuels and bioprocessing
- Cryogenic separation
- Bulk terminals
- Oil and gas pipeline
- Transportation and bulk shipping
- Weapons manufacture
- Water treatment
- Waste treatment

Often, supervisory roles are in the best position to drive change both up and down in the organizational hierarchy. Such change will require strong support from the organization's senior leaders. We want to help you understand the concepts in order to influence players at all levels of the organization. The last chapter of the book, Chapter 12, *A Call to Action,* provides some guidance in achieving this goal through risk communication and individual empowerment. A person detecting a warning sign and failing to present a credible case to management for how to address it leaves an organization vulnerable to failure. Future efforts to communicate after a failed attempt may not go well. If the need for action and improvement is not universally recognized, nothing will change.

Anyone interested in understanding, identifying, and taking action related to catastrophic incident warning signs can benefit from this book. The following list suggests personnel who may reap an advantage from this book.

- Operations, maintenance, and other manufacturing personnel who may be critical in implementing parts of a facility's process safety management system
- Safety professionals at the facility and corporate levels
- Process safety and risk management program managers and coordinators at manufacturing facilities
- Corporate process safety management staff
- Project managers and project team members whose projects have process safety implications
- Engineers or other staff members initiating changes under the management of change protocol
- Facility managers and other members of manufacturing plant management teams responsible for the overall safety of chemical processing facilities
- Regulatory agency staff involved in the initial permitting and ongoing compliance of processing facilities with relevant process safety standards

The concepts in this book are helpful to all who strive to develop a culture where safety is exhibited in everyday actions.

1.7   HOW TO USE THIS BOOK

Each of the chapters on warning signs provides a brief overview of the related management systems, associated warning signs, and a relevant case study. This book provides you with an opportunity to evaluate the warning signs, their associated process safety elements, and the case studies of incidents in which the warning signs may have been present as precursors.

If you find that several warning signs exist within a single element category at your facility, thoroughly check all the associated warning signs. If you cannot readily correct the ones you find, you may want to conduct a thorough audit. This can help identify appropriate follow-up actions. Formal audit findings can address some of the prevalent warning signs. This book can also help you strengthen an organization's audit protocol and provide a cross-check against a process safety audit report.

Finding numerous warning signs across multiple categories may signal the need for significant improvements in the facility's process safety culture. The Baker Panel Report issued in January 2007 as part of the follow-up to the 2005 BP Texas City explosion provides guidance for safety culture improvement.

Finally, this book is useful for safety meeting topics. The leader can refer to examples of past events and ask the questions—*What if*—and—*Could this happen here?* A reference to past incidents is particularly valuable in today's operating environment. Many of the workers and leaders who were involved in your site's process safety management system's initial implementation are leaving the industry. The rate of experience loss in the process industries is expected to increase over the next several years. Without some guidance, today's staff may not realize that they are on a path toward repeating past incidents.

1.8   CASE STUDY – TOXIC GAS RELEASE IN INDIA

The 1984 Bhopal, India, incident ranks as the worst chemical facility disaster in history. It claimed the lives of 3000 to 10,000 people and injured 100,000 others. An analysis of the Bhopal incident reveals failures in each element of process safety. For this reason, we have used the Bhopal case study to support our introductory chapter. It occurred on an average night when work practices and habits were typical of day-to-day management.

The incident took place in an intermediate storage area of the facility where liquid methyl isocyanate (MIC) was stored in three separate tanks embedded in a berm. MIC is a feedstock used in the production of carbamate, a common insecticide. MIC liquid is highly reactive in the presence of water and iron oxide, and it generates heat. In sufficient quantities, this heat may generate highly toxic vapor. The process design included a refrigeration coil to ensure that the temperature could not exceed 5°C and a vent gas scrubber to prevent vapor escape. In addition, despite a low process operating pressure, a closed relief and blowdown subsystem was installed to further decrease the risk.

For several months before the catastrophic incident, conditions at the facility had been deteriorating. For example, workers did not follow procedures carefully and several mechanical features were either shut down or compromised. Examples include the refrigeration circuit, which was low on coolant, and the vent gas scrubber and flare system, which was out-of-service. The temperature indicator on one tank was defective and the temperature in another of the tanks exceeded the maximum limit by as much as 15°C with no corrective action.

On the night of the catastrophic incident, operators heard a screeching noise from the relief valve on one of the tanks. Unfortunately, the closed blowdown system was out of service for maintenance. At the time, workers were cleaning out a clogged pipe with water about 400 feet from the tank. It was later thought that while operators were on their shift change or on a break, someone could have intentionally disconnected a pressure gauge from the cover plate on one of the tanks and attached a water hose. A significant quantity of water entered one of the tanks and triggered a runaway reaction and subsequent release of MIC vapor into the community. Another theory suggested that due to ineffective maintenance and leaking valves, water may have leaked into the tank.

Principal factors leading to the magnitude of the gas leak include the following items.

- Storing MIC in large tanks and filling beyond recommended levels
- Poor maintenance
- Failure of several safety systems (due to poor maintenance)

- Safety systems being switched off to save money—including the MIC tank refrigeration system, which could have mitigated the disaster severity

Other physical contributors to the incident included the following items.

- The MIC tank alarms had not worked for at least four years.
- There was only one manual backup system, compared to a four-stage system used at facilities operated by the parent company.
- The flare tower and the vent gas scrubber had been out of service for five months before the disaster. The gas scrubber, therefore, did not treat escaping gases with sodium hydroxide (caustic soda), which might have brought the concentration down to a safe level, although the scrubber and flare system design was inadequate for the flow rate of MIC involved in the incident.
- To reduce energy costs, the refrigeration system, designed to inhibit the escape of MIC vapor, had been left idle; the MIC was stored at a hotter temperature than recommended. It was determined that the refrigerant was being used to cool the management offices.
- The steam boiler, intended to clean the pipes, was out of service for unknown reasons.
- Pipe blinds that would have prevented water from pipes being cleaned from leaking into the MIC tanks through faulty valves were not installed.
- The water pressure was too low to spray the escaping gases from the stack. They could not spray high enough to reduce the concentration of escaping gas.
- According to facility operators, the pressure gauge on the MIC tank had been malfunctioning for one week prior to the incident.
- Carbon steel valves were widely used at the facility, even though they corrode when exposed to acid.

The technical details provided above were not made available to the investigators for a period of one year. A lockout by the Indian government prevented technical experts from the parent company from entering the site. An analysis of chemical residue in various pieces of process equipment was used to support the development of incident logic.

The incident at Bhopal was not really an accident. There was no real element of surprise. Conditions had deteriorated for many months while no one stepped forward to intervene. Physical and paper evidence suggesting huge gaps in the safety culture were very visible and should have been apparent to any responsible worker on the facility site. However, the definition of responsible worker has little meaning if management has not defined a safety culture and declared a clear set of operating priorities. Process safety culture is a state of excellence that starts with strong leadership and demands discipline and accountability at all levels.

Several additional questions need to be asked.

- Why did the parent company not enforce its own policies and procedures?
- Why did workers not complain about conditions in the facility?
- Why was defective equipment not properly maintained and repaired?
- Why were emergency plans and procedures not developed?
- Why was there no dialogue between facility management and the local community that might have led to a timely evacuation?
- Why was there little cooperation between the company and the government in accessing evidence and reaching timely conclusions?

This catastrophic incident led to the financial ruin of the parent company and resulted in the development of process safety protocols and regulations worldwide.

Exercise: Some of the warning signs that may have been noticed prior to this incident are listed below.

- Operating outside the safe operating envelope is accepted.
- Operation continues when safeguards are known to be impaired.
- Critical safety systems are not functioning properly or are not tested.

Can you identify other warning signs that may have been present?

2
INCIDENT MECHANICS

There are no accidents without intentions.
Alex Miller

2.1 INCIDENTS DO NOT JUST HAPPEN

Catastrophic incidents do not just happen. They often result from a fundamental weakness (or weaknesses) in the management systems used to control an operation. In many cases, these weaknesses have existed for a long time.

Catastrophic incidents are undesirable events that can result in serious injuries, deaths, environmental damage, and significant business losses. In the process industries, these incidents typically involve the release of harmful materials from systems designed to contain them. When these releases occur, employees, contractors, and members of the public could be seriously harmed. Ultimately, such catastrophic incidents can put a company out of business. The quote below describes the reality of accidents:

Most accidents are preventable.
Planes do not fall out of the sky unless something is wrong.
***Mary Schiavo, former inspector general of United States Department of
Transportation***

If we want to avoid catastrophic incidents, we need to make an effort to pinpoint their origins and take action while the opportunity exists to fix what is wrong. To achieve this goal requires a high level of commitment and an understanding of both how large-scale incidents occur and the warning signs that precede them.

2.2 INCIDENT MODELS

Several models of incident causation exist. Each can help us better interpret the warning signs that precede incidents and help in reducing their likelihood and impact.

Incident investigators use their judgment to make adaptations to selected techniques to drive the size and complexity of the investigation effort. They identify pertinent facts and apply principles of logic and reasoning skills to determine how and why an incident occurred.

2.2.1   The difference between incidents and catastrophic incidents

The terms *incident* and *catastrophic incident* can be confusing and may result in misunderstanding. We define the terms as follows:

- An *incident* is any unplanned or undesired event either with or without consequence. A near miss is an incident with no measurable consequences.
- A *catastrophic incident*, sometimes referred to as a catastrophic accident, is a large-scale event with high-level consequences. Such consequences may include serious injuries or fatalities and widespread damage to facility and equipment.

Catastrophic incidents are rarely the result of a single cause. They are typically a culmination of minor incidents that occurred over time. The time sequence of a catastrophic incident may be short lived, or it may have developed over several weeks, months, or years. In either case, the management system deficiencies that contributed to the catastrophic incident were often present long before the catastrophe.

In the words of Trevor Kletz, a process safety expert: "If you think safety is expensive, try having an accident. Accidents cost a lot of money in terms of damage to the plant, claims for injuries, and the loss of a company's reputation."

The Safety Pyramid (Figure 2-1) illustrates various metrics to manage to help prevent process safety incidents. The pyramid model illustrates the relative numerical relationship between catastrophic incidents and substandard conditions. Those events with little or no impact at the bottom of the pyramid far outnumber the major incidents seen at the top. This statistic may provide a false sense of security for those uninitiated in the field of safety. The fact is that each of the lower level incidents had the potential to escalate had it not been for some preventative or mitigating activity or barrier. As safety barriers degrade and substandard practices take hold, the potential for catastrophic incidents increases in proportion to the number of near misses.

Figure 2-1   Safety Pyramid

Some experts suggest that organizations should focus most of their attention on avoiding catastrophic incidents rather than spending effort on incidents that cause little harm or consequence. With the global trend toward cost reduction and the current shortage of capital in many industries, this approach might appear to make good business sense. However, the reality is that no one knows how to distinguish the warning signs of less-than-catastrophic incidents from those of catastrophic incidents. Remember that catastrophic incidents are incidents with significantly elevated consequences that develop from less consequential incidents. If we delay our response until an incident cluster has escalated in consequence potential, it may be too late to take effective action. Additionally, the selling point for a strong safety program should not be based solely upon avoiding catastrophic incidents. A series of well-publicized minor events could just as easily affect an organization's reputation and viability.

2.2.2   The Swiss cheese incident model

For many years, safety experts have used the familiar Swiss cheese model (after Reason, 1990, as adapted by Weigmann & Shappell, 1999) as one theory to explain catastrophic incidents.

The model can help managers and workers in process industries understand the events, failures, and decisions that can cause an incident or near miss to occur. The example in Figure 2-2 depicts layers of protection that all have holes. When a set of unique circumstances occur, the holes line up and allow an incident to happen. We display protection layers in the figure as slices of cheese. The holes in the cheese represent the following potential failures in the protection layers:

- Human errors
- Management decisions
- Single-point equipment failures or malfunctions
- Knowledge deficiencies
- Management system inadequacies, such as a failure to perform hazard analyses, failure to recognize and manage changes, or inadequate follow-up on previously experienced incident warning signs

As the figure illustrates, incidents are typically the result of multiple failures to address hazards effectively. The model also recognizes that a catastrophic incident is a vector, a moving force, which is fueled by multiplying failures that deepen its effect as it progresses. Ultimately, the impact can harm people, assets, and the environment. Management systems are set in place to help obstruct the trajectory or block its path. A management system may include physical safety devices or planned activities that protect and guard against failure. Process safety management systems are more than just plans. They are a business philosophy described by comprehensive strategies supported with live actions that continue relentlessly even when the results appear good. An effective process safety management system can reduce the number of holes and the size of the holes in each of the systems layers.

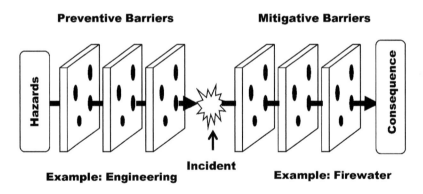

Figure 2-2   Swiss Cheese Model of Barriers to Consequences

Management system elements serve several distinct purposes. Some deal with people while others address the equipment integrity within a facility. Still others check the integrity of the process to ensure that it is stable and predictable. Within a company, there may also be management systems that deal with financial results and stakeholder relations. Some management systems are driven by the corporation and apply to several facilities. Others are facility specific. From this point forward, the book concentrates on those systems that affect process safety at a facility.

Each process safety management system element serves to prevent incidents from occurring. Sometimes the element's effectiveness level is overlooked or simply assumed to be strong enough. The elements serve as preventive layers (or barriers to consequence) in the incident model. Examples include the following items:

- Appropriate process design
- High-quality construction
- Well-written procedures
- A well-trained workforce
- Operational discipline
- Performance management
- A quality assurance program that addresses equipment integrity and reliability
- An established management of change process
- An established engineering design process
- Systems for communicating lessons learned from incidents
- A system for conducting audits

Elements that prevent or reduce the impact of an incident are *mitigative layers*. Examples of mitigative layers include safety barriers, containment walls, drainage ditches, relief valves, fire suppression systems, and emergency planning. At the end of the list, emergency planning, we find the management systems that deal with incident response and recovery. Successfully implementing well-planned incident response and recovery operations may be important for an organization's short and long-term viability and are sometimes overlooked.

In the real world, management systems are never perfect. They might be viewed as software for humans. Moreover, humans are fallible even when that software is accurate and effective. Commitments may be broken. People may sometimes neglect or be distracted from their responsibilities, especially when the consequences of doing so are not readily apparent. For example, a procedure that needs to be updated today as a result of a change in the field may be deferred until later in the month after startup while other work of a more desirable or urgent nature is done. These cracks or imperfections represent weaknesses in the systems used to manage an operation. The Swiss cheese model shows these cracks or imperfections as holes that are penetrable by the incident event chain. The size of the holes and their relative number vary with the nature and frequency of the imperfections. Weak systems are more prone to failure and are least effective at preventing incidents from occurring or mitigating them once they occur.

2.2.3   The bonfire incident analogy

The opportunities for prevention are most prevalent at the bottom of the incident pyramid. This is especially true for companies that have never experienced a catastrophic incident. System weaknesses and deficiencies are like slow-burning embers at the base of a small fire (Figure 2-3). As the timbers ignite, the fire spreads rapidly, making it far more difficult to control. If an organization is to improve its safety performance, changes need to address the existing gaps and deficiencies.

Figure 2-3   Remove the Fuel

2.2.4   The dam incident analogy

The analogy of the accumulation of water behind a dam (Figure 2-4) is another example that illustrates how catastrophic incidents occur. As the water level reaches the top, the hydraulic head approaches the limits of the dam's design and construction specifications. Any small leaks encountered during the filling operation signal the need to discontinue adding more water or to make immediate repairs. If failure occurs at the water's high level, the results are often catastrophic. It is impossible to reverse failures and counteract momentum to prevent a major event.

Warning signs may only be present periodically and for a brief time. When we notice a warning sign, it is important to take corrective action immediately. If a warning sign reappears occasionally, the organization should identify the applicable management system weaknesses and take corrective action until the preventative systems are in order. Liken this to a toothache or medical problem that appears to worsen at certain times. Such a problem, if overlooked, will certainly recur.

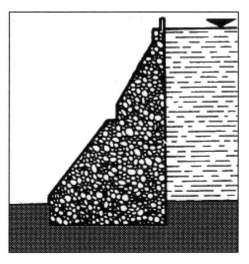

Figure 2-4   The Dam Model

2.2.5   The iceberg incident analogy

Another concept commonly applied is the Iceberg Model (Figure 2-5). While only 10% of an iceberg is visible above water, the large mass below the surface causes the most damage in a collision. Things we do not see often hurt us the most. In an industrial setting, we train workers to deal with physical and chemical hazards. The physical hazards are usually more noticeable than the chemical ones. We can detect hazards visually or by the use of specialized equipment. Systemic weaknesses, on the other hand, are less tangible. If we do not detect and address them, they can be far more significant. Consequently, any effort to improve the management systems will be far more effective than merely treating the physical symptoms. To support this point, Chapter 11, *Physical Warning Signs,* is located near the end of this book so that it does not overshadow the other chapters. In order to prevent losses from occurring, strike a balance between addressing the physical warning signs and the less obvious management system related warning signs.

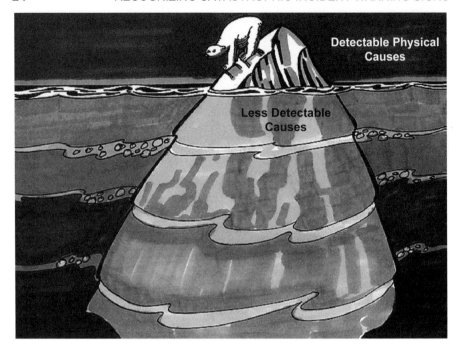

Figure 2-5   The Iceberg Model (Image AEI 2002)

2.2.6   Incident trends and statistics

Incident trending and statistical analysis provides a powerful tool for portraying deteriorating conditions in a facility. It often suggests that a major incident may be imminent. In its simplest form, a monthly trend analysis can depict whether an indicator of deteriorating conditions is increasing or remaining constant. Do not confuse occupational safety trends with process safety related trends.

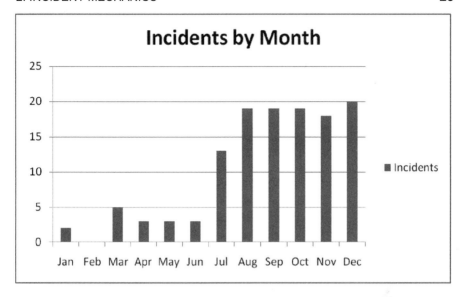

Figure 2-6   Incident Trends

The response to a trend plot such as that in Figure 2-6 should be to determine what other changes might have occurred during the same time span. This information can serve to reinforce the message that conditions are deteriorating. Repeat events can be viewed negatively by regulators, communities, and shareholders, and can lead to adverse business effects. A repeat event signifies organizational issues over and above those likely to have been identified in a single incident.

Near miss reporting provides an excellent opportunity for studying the process safety culture of an organization. Very often, the difference between a near miss and a catastrophic incident is mere luck. Using the Swiss cheese model shown in Figure 2-2 to analyze a near miss, the precise conditions for maximum loss did not line up. If we allow similar near misses to occur without implementing controls, eventually the holes in the Swiss cheese will line up. When this happens, a catastrophic incident will result.

An analysis of minor incidents and near miss reports should examine the type of scenario, the time, the location within the facility, the experience of the workers, and any causal or mitigative factors identified. These analyses should attempt to extract common themes which can then be addressed through follow-up.

2.2.7   Root cause analysis

Root cause analysis (RCA) is a top-down analysis of the events and causal factors that contributed to an incident. RCA often makes use of logic trees to arrange the data describing an incident into a communicable and documentable form, as shown in Figure 2-7. Principles of necessity and sufficiency checking are applied to the logic described in the trees to facilitate quality review of both data and reasoning. This characterizes root cause analysis as an iterative process. The method is well suited to work teams. RCA can be applied as a tool for supporting continuous improvement to management systems.

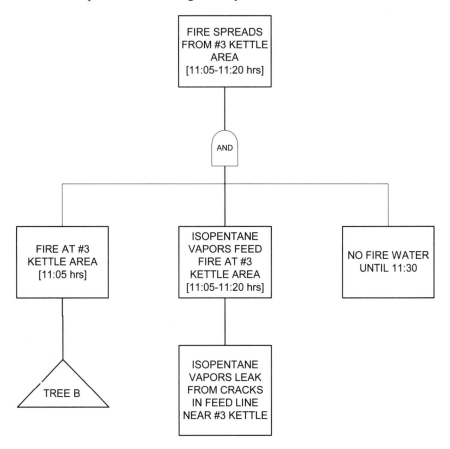

Figure 2-7   Simplified Logic Tree Root Cause Analysis

2.2.8   Multiple root cause theory

For a very serious incident to occur, failures often will have occurred at several levels and along several fronts. If one management system element is very weak, others are probably weak as well. Identifying a single root cause can lead to inappropriate preventive and corrective controls.

The multiple root cause theory suggests that we need to examine and address parallel weaknesses to prevent significant incidents. A root cause is often not the absence of a management system but a failure within the management system.

2.3   CASE STUDY – BENZENE PLANT EXPLOSION IN CHINA

In late 2005, a nitrobenzene distillation column exploded in a benzene production facility in Jilin, China, causing eight deaths and 60 injuries. The direct economic loss was ¥69.08 million RMB (Renminbi) or $10,000,000 USD. The pollutant waste streams from the incident flowed into the Song Hua Jiang (White River), causing acute water pollution.

The direct cause of the explosion was an operating error. A process operator in the nitrobenzene facility had incorrectly executed an operating procedure during a critical phase of operation. When the operator stopped raw nitrobenzene feed to the column, he did not close the steam valve of the preheater. This led to the vaporization of the material inside the preheater. In the process of restarting the nitrobenzene unit, he wrongly executed the operating procedure once again. He opened the preheater's steam valve to heat up the system, and then started the raw nitrobenzene feed pump. This caused violent vibration of the material fed into the preheater. The vibration loosened the preheater flanges and connected piping. This condition caused failure of the purge system and allowed air into the process. The distillation column and other devices exploded (see Figure 2-8) and caused catastrophic damage to the facility.

A root cause analysis of the explosion determined that management did not pay much attention to process safety. Specifically, there was little evidence to suggest that workers were required to follow procedures. This incident escalated over a prolonged period. There may have been several opportunities to correct the course of events.

The direct cause of the pollution aspect of the incident was that measures to prevent the polluted water from flowing into the river were not available in the facility. After the explosion, ineffective mitigation actions (weak emergency planning and response) allowed the polluted water to flow into the Song Hua Jiang.

The root cause of this significant event was that corporate and facility management had not considered the possible consequences of major incidents resulting from failure to enforce process safety protocols. The emergency plan was inadequate and the company management team did not put environmental protection in the appropriate perspective. Meanwhile the local emergency rescue headquarters underestimated the consequence and paid inadequate attention to the water pollution. Mitigating measures and protective requirements were not put in place in a timely fashion.

Figure 2-8   The Jilin Explosion Aftermath (AP Photo 2005)

This catastrophic incident was clearly the result of poor operational discipline, an insufficient and ineffective process safety management system, and an inadequate emergency response plan. In turn, these reflect weak process safety leadership. It is unknown whether previous incidents of a similar nature had been experienced with or without consequences. A company's past is often an indication of future events. Strong management systems should have prevented and detected the initial problems that led to this incident. Clearly defined responsibilities would have ensured that workers were prepared to respond to the unexpected.

Exercise: Can you identify warning signs that may have preceded this incident?

3
LEADERSHIP AND CULTURE

If your actions inspire others to dream more, learn more, do more,
and become more, you are a leader.
John Quincy Adams

3.1 HOW DOES LEADERSHIP AFFECT CULTURE?

Leadership is the ability to influence positive change and guide other people toward successful outcomes. Some behavioral characteristics of leadership are inherent in some people, but many of the competencies associated with excellent leadership are teachable. All effective leaders benefit from a combination of experience and training.

One characteristic of a good leader is open communications. Leaders need to listen to their subordinates and peers. When a worker observes something that is unacceptable or outside the norm, he or she should feel encouraged to speak out or address the problem without fear of reprisal.

Another characteristic of a good leader is accountability. Leaders need to hold themselves and their subordinates accountable for assigned work and associated goals. A responsible leader provides the tools and resources to help ensure that the team can perform work safely and will remove obstacles when they occur. Responsible leaders do not hide behind excuses or deny the presence or importance of incident warning signs. They look for opportunities to improve situations.

If a leader stays in close communication with the workforce, there should be few surprises and less of a need to react. Leadership does not reside solely in the formal chain of command as defined within a facility's organization chart. Ultimately, all workers should assume a level of leadership within their teams when called upon in order for a process safety system to be effective. Good leadership is the ability to empower employees, not micromanage them.

3.1.1   Communication

Effective communication between workers and among working groups is critical. Communication must be vertical as well as horizontal. This means that workers must communicate up through the formal chain of command as well as among themselves. In his books and lectures, Professor Andrew Hopkins has emphasized the importance of written and verbal communications regarding unsafe or unusual conditions. These communications need to be formal, as with shift reports, as well as informal and need to involve all levels of workers.

One of the most common warning signs of failing communications is the development of an *us against them* mentality. This can occur between shifts, between departments, or between other areas of the facility. In any organization, relationships between working groups can become difficult for many reasons. An effort may be required to get communications back on track.

One of the most threatening situations is that of a strained relationship between management and hourly personnel. Strained communication between these groups is sometimes a result of a failure on management's part for not seeing the early warning signs and initiating constructive, effective communication. Make sure that everyone sees the management team as a body that is willing to hear bad news and act prudently upon that news. The physical presence of management personnel to see operating practices and work conditions helps reinforce good communication.

3.1.2   Operational discipline

Operational discipline is displaying behaviors within a system of checks and balances that help ensure that things are done correctly and consistently. A high level of operational discipline supports consistent success.

Examples of behaviors that demonstrate a high level of operational discipline include the following:

- Consistent practice of established work processes and procedures
- Effective shift turnover practices
- Consistent use of safe work permits to control work
- Effective and consistent use of interlocks
- Consistent use of bonding and grounding practices
- Excellent general housekeeping
- Consistent use of personal protective equipment
- Consistently applied security measures

In highly hazardous situations, you may need to develop and employ additional controls. Operational discipline is a subcomponent of the RBPS (Risk Based Process Safety) element *conduct of operations*.

3.1.3   Process safety culture

Organizational culture is often described as *the way we behave when no one is watching*. Process safety culture is a subset of an organization's overall culture. Process safety culture arises from a common set of values, behaviors, and norms that, when displayed, affects process safety performance. A facility with an effective process safety culture will display an excellent process safety record and is more likely to recognize and attend to catastrophic incident warning signs when they appear. A facility with a weak process safety culture may see numerous warning signs that are allowed to exist for extended periods.

Symptoms of weak process safety culture include the following:

- Complex and unclear responsibilities and accountabilities
- Isolated functional groups within the organization
- Poor communication within the plant or organization

The symptom of a strong process safety culture is not perfection. It is timely progress made on process safety related action items. If a site with a positive process safety culture seems to be lagging behind for some reason, consider whether the following situations exist:

- The action items were not adequately defined or technically well founded.
- The action items include complex work that takes time to complete.
- The person or group that was given follow-up responsibility did not understand the action items.
- The plant or facility does not have the resources or skills to pursue the action items.
- Local site management does not promote action item follow-up as a high priority.
- The action item was not adequately funded or funds were used to address other facility needs.

The concepts of operational discipline and process safety culture form a positive feedback loop. A high level of operational discipline supports an effective process safety culture, and an effective process safety culture supports a high level of operational discipline. In contrast, a low level of operational discipline is generally found at a facility with a weak process safety culture, and a weak process safety culture supports a low level of operational discipline. A good leader and an effective leadership team will be alert to slipping process safety culture symptoms and low levels of operational discipline.

3.1.4   Process safety versus occupational safety

The goal of occupational safety is the avoidance of physical injuries such as slips, trips, falls, strains, chemical exposure, and electrocution. Such incidents are usually associated with a hazard that is close to the worker. Occupational safety focuses on the individual and assumes that, given a well-designed safety management system, proper training, procedures, tools, and protective equipment, a worker can perform his or her work safely. Immediate control of this type of work resides primarily with the individual.

Process safety requires a systematic approach to managing hazardous processes and operations that have the potential to release harmful materials with catastrophic consequences. Safety can only be assured if all systems are functioning in harmony. Process safety relates to the quantity, quality, and variety of controls or protective features that protect people, the environment, and property from process hazards. Process safety requires the commitment and involvement of all workers, including management and contractors. At stake are the lives of workers as well as members of the public and the environment.

Injury statistics have traditionally been used to gauge the effectiveness of safety measures on a worksite. Compilation of such statistics is a regulatory requirement in many jurisdictions. Injury rates are seldom the best indicators of process safety performance. Leading indicators that reflect current activity provide a more accurate indication of the commitment to process safety within an organization.

Unless leaders clearly understand the difference between occupational safety and process safety and communicate the expectations, confusion will exist and process safety performance will suffer.

3.2   THE LEADERSHIP AND CULTURE RELATED WARNING SIGNS

The warning signs associated with leadership and culture are provided below.

- Operating outside the safe operating envelope is accepted
- Job roles and responsibilities not well defined, confusing, or unclear
- Negative external complaints
- Signs of worker fatigue
- Widespread confusion between occupational safety and process safety
- Frequent organizational changes
- Conflict between production goals and safety goals
- Process safety budget reduced
- Strained communications between management and workers
- Overdue process safety action items
- Slow management response to process safety concerns
- A perception that management does not listen
- A lack of trust in field supervision
- Employee opinion surveys give negative feedback
- Leadership behavior implies that public reputation is more important than process safety
- Conflicting job priorities

- Everyone is too busy
- Frequent changes in priorities
- Conflict between workers and management concerning working conditions
- Leaders obviously value activity-based behavior over outcome-based behavior
- Inappropriate supervisory behavior
- Supervisors and leaders not formally prepared for management roles
- A poorly defined chain of command
- Workers not aware of or not committed to standards
- Favoritism exists in the organization
- A high absenteeism rate
- An employee turnover issue exists
- Varying shift team operating practices and protocols
- Frequent changes in ownership

3.2.1 Operating outside the safe operating envelope is accepted

The safe operating envelope (SOE) is a zone in which changes and adjustments may be made to a process in accordance with normal operating procedures with no anticipated adverse effects. The SOE for any process is typically defined in terms of temperature, flow, pressure, and composition. Process equipment is designed on the assumption that a process will be operated within its SOE throughout its operating cycle. When a process drifts outside the envelope, the risk of a serious incident may increase significantly. Process equipment might experience unintended conditions, and a mechanical failure could result. At elevated conditions, the process materials may react more violently and could experience a runaway reaction. Operating outside the safe operating envelope for any reason is unsafe and puts the entire operation at risk.

Tolerance for routine excursions outside the SOE suggests a major failure in leadership and operational discipline. It is imperative that all workers understand the importance of controlling the operation within allowable limits and be accountable for ensuring that breaches do not occur. Many operations post the safe operating limits as a constant reminder of this requirement.

Another concern possibly related to this warning sign is that leadership may not fully understand the importance of adhering to SOE limits and the associated risk of noncompliance. Leadership can only be effective through a thorough understanding of hazards and risks at the field level. Some typical questions for leadership are provided below.

- Is a safe operating envelope defined for all operations and processes?
- Is it reflected in operating procedures and training manuals?
- Are the consequences of deviating from the safe operating envelope clear to all workers?

- Does facility management stress the importance of controlling the operation within the safe operating envelope?
- Are deviations from the safe operating envelope investigated?

3.2.2   Job roles and responsibilities not well defined, confusing, or unclear

By developing specific task lists for each position at a site, an organization exhibits a high level of operational discipline to both teams and individuals. Job roles and responsibilities are well defined and clear.

The operation of a processing facility is often complex and involves multiple processes and equipment. Tasks and procedures need to be carefully coordinated to avoid incidents. We plan for some special tasks. We conduct other tasks on a routine basis. Still other tasks are performed in response to upset or abnormal operating conditions. Coordination of roles and responsibilities helps ensure that we allocate all tasks in terms of content, quality, and timing (correct sequence) without duplication or overlap.

Without coordination, workers can make incorrect judgments in the field. A well-intended worker may inadvertently drain a line that was just placed in service by another worker. It is possible to omit, overlook, or perform important tasks in the wrong order. The result of improper job coordination can be human error, which has been a contributing factor in many serious incidents. Effective field communications are important in the operation of any facility. However, they cannot replace effective leadership. Some questions to consider are provided below.

- Are job roles and responsibilities well defined, or are they confusing, ambiguous, or unclear?
- Are job responsibilities clear to all workers?
- Is there a full facility curriculum for each job position—manager to laborer—addressing each process safety related task assigned to that specific job position?
- Are workers trained adequately to handle their assigned responsibilities?
- Is a mechanism in place to request additional help when workers cannot handle an assigned job?
- Is a supervisor or mentor available to address concerns related to job assignments in the field?

3.2.3   Negative external complaints

When negative complaints are received from the community, something may have changed beyond the knowledge of those in the plant or facility. Do negative reports cause a reevaluation of your business operations, or does the organization consider them fringe comments and ignore them? The community may question the organization's performance in being socially responsible.

- Are you taking positive, measurable actions to resolve the findings?
- Do you need to improve your public relations with the community?
- Do you make open and complete reports on process safety performance available through the company website or through press releases?

3.2.4   Signs of worker fatigue

It is generally expected that workers will report to their jobs in good physical health prepared to perform their work. Occupational health and safety regulations in many areas define the requirements for workers to be physically fit to do work. With extended hours during shift work (twelve-hour shifts plus commuting time), there is a possibility workers will become emotionally and physically fatigued toward the end of a long shift. The problem may be further complicated by long commuting distances to return home between shifts. This reduces the rest time before returning to work. Another contributor to worker fatigue is extended work assignments involving overtime during shutdowns or special projects.

Workers themselves often place pressure on their peers to work excessive overtime. Studies show that workers are far less capable of performing their jobs safely when they are fatigued. When personnel are working too many hours, it is difficult to ensure that the workforce will follow work processes and procedures. Tired workers are less likely to spot hazards and take appropriate actions. Organizations need to examine staffing levels and make sure that there are adequate resources to perform the work. Avoid scheduling critical jobs near the end of a shift, on Friday afternoon, or during night hours. If this warning sign persists, an audit of work practices in the field may be needed to better define the problem and determine a solution.

- Do workers at your facility appear to lack energy to perform their jobs?
- Is the percentage of overtime worked by certain departments or individuals high?
- Do workers appear demotivated at certain times?
- Are there restrictions on the number of hours worked on a weekly or monthly basis?
- Are facilities available for workers to take exercise during lunch breaks or time off?
- Are workers encouraged to rest and relax during their time off?

3.2.5   Widespread confusion between occupational safety and process safety

This warning sign suggests that your facility employees—plant management, staff, and hourly personnel alike—do not understand process safety. Process safety shares a common set of values with occupational safety; that is, it protects workers from harm. Unless we understand the differences, neither will be managed effectively.

This warning sign is common even in high hazard industries. A single manager, untrained in process safety, can unintentionally make an operating decision inconsistent with the principles of process safety. The Baker panel report following the 2005 BP Texas City refinery explosion clearly cited this warning sign as a significant contributor to the chain of events that led to the accident. The panel also warned of similar confusion at other facilities in the oil-refining sector. Which of the following statements might be true in your organization?

- Injury rates are the most often used key performance indicator (KPI) for managing and measuring safety performance.
- Company communications do not use the phrase *process safety* very often.
- Minor loss of containment incidents are sometimes not reported or investigated.

3.2.6   Frequent organizational changes

Frequent organizational changes can introduce confusion into an organization and may release certain individuals from process safety follow-up commitments. It is common to see cases where there are frequent movements of key people due to promotion or other reasons. People can lose track of critical things during interim periods related to organizational change.

Organizational changes are a normal part of running a business. Promoting workers to a higher position recognizes their skills and efforts. The expectation is that such individuals will be able to function equally well in the new position. Training must occur for this to happen. An organizational change should be analyzed for its impact on process safety. It will require each worker to adjust or alter his or her personal mode of communication with the incumbent. An organizational change can leave vulnerable gaps in communication if handled improperly. Previous promises or commitments may need to be revisited in light of new responsibilities. Ideally, a period of overlap or transition is planned to allow newly assigned workers to work alongside their predecessors.

In some cases, organizational changes occur due to illness or sudden departure of workers in key positions. Succession plans can help address such anomalies. However, carefully manage and plan organization changes when possible.

Frequent organizational changes can compromise stability and continuity. Rapid movement of personnel can challenge the ability of assigned persons to learn a new job and contribute to the effectiveness of the operation. Frequent organizational changes also reduce accountability.

- Are there trigger events to initiate management of change whenever a key person within the organization is transferred?
- Do you include this trigger in your management of change system?
- Has your organization developed tools to assist in transferring the process safety responsibilities and tasks to the new owners?

- Are succession plans in place to address the loss of key persons within the organization?

3.2.7   Conflict between production goals and safety goals

This organizational behavior represents the attitude—*pounds, barrels, or gallons out the door*—as being paramount to personnel and process safety interests. This behavior does not support people sensing a high level of organizational operational discipline. The business case for process safety clearly demonstrates that process safety excellence achieves a higher level of plant availability and throughput. Leaders with short-term views may often overlook this reality.

- Do you commend workers who are willing to stop production to prevent increased risk?
- Do you ensure that their good decision-making skills are recognized within the group?

3.2.8   Process safety budget reduced

If process safety budget cuts occur without a fully accountable reason, it often signals that the organization is becoming less aware of the value of embracing process safety as a part of its business philosophy. Companies have seen firsthand that it is impossible to maintain a high-quality management system with a minimized process safety staff or budget. Process safety is costly; planning and budgeting should be in place to ensure that cost reductions are not imposed arbitrarily.

- Does your facility have a process safety budget?
- Is the process safety budget being spent effectively?
- Is the process safety budget (and associated expenditures) discussed in the plant leadership meetings on a regular basis?
- Is a mechanism in place to request additional funding to address high-risk situations or changes in the PSM plan?

3.2.9   Strained communications between management and workers

Strained communications between leaders and workers can undermine the morale of the workforce. This can affect the commitment of workers as well as the quality of work. Ultimately, errors and shortcuts may become commonplace. Management should make an effort to demonstrate improved communication skills consistently, and the organization should train all employees in the concepts of effective communication techniques.

- Is management presence visible in the field?
- Has management attempted to talk directly to workers on the front line?
- Do the lines of communication seem strained between the organization's leadership and the hourly staff?

- Is the perception of the hourly employees that management does not listen to or act on their issues of concern?
- Is there unusual friction between members of the same organization? For example, are there problems between shifts in operations, between trades in maintenance, or between other groups?
- Are the relations between different working groups strained? For example, are there issues between operations and maintenance, or between the inspection, and engineering groups?
- Are there adequate lines of open communication between these groups? If not, what other communication opportunities are there?

- If there are strained communications between groups, do you know what each group's main points of contention are? If not, are you listening enough?
- Are there cliques within your organization, between shifts, departments, or even employees versus contractors?

3.2.10 Overdue process safety action items

Process safety activities generate follow-up activities for assignment to individuals or groups based on their knowledge and expertise. Many factors might determine the time required to complete and close out an assigned action. A disciplined management system that considers priority, scope, and resources will help ensure that important jobs are completed when required. The challenge is not simply completing an assigned list, but prioritizing the action items on the list. Action item closeout documentation should provide the rationale and conclusions.

Even if we prioritize the action items, a large collection of incomplete low-priority action items can also result in a significant loss event. Therefore, as we identify and prioritize the actions, we need to take the bold action of removing the low-priority (low risk) items from the list completely. If we remove the clutter (which is what very low-priority action items really are), we can better concentrate limited resources on the vital few that, if not done, can have a significant effect on process safety.

- Are your action item lists growing in size?
- Have you prioritized every action item in the list?
- Have you made a conscious effort to complete the simplest items?
- Do you evaluate, and possibly remove, the lowest prioritized actions?
- Does your organization need to evaluate its system for establishing priorities?
- Are action item completion dates set within a realistic period?
- Are process safety related items prioritized using a risk-based approach?
- Does your organization review the action item list on a regular basis?

If you have significant backlogs of overdue work items identified to support any driver—personnel safety, process safety, environmental compliance, quality, or economics related—personnel might be at risk.

When companies recognize this issue, sometimes the blitz approach used to solve it results in activity-based closure rather than outcome-based closure to satisfy the driver for the action item.

3.2.11 Slow management response to process safety concerns

When staff members at a facility develop this perception, it may undermine the process safety culture. The idea that—*if management does not care, why should we?*—may spread across the organization. Senior managers are direct stakeholders in process safety. Management should put catastrophic incident prevention at the top of its priorities.

- Are process safety concerns that require management follow-up logged, and is the status regularly communicated to the workers?
- Is an individual designated to communicate with management on a regular basis for updates?
- Are reported issues dealt with within a reasonable period?

3.2.12 A perception that management does not listen

When management openly discounts employee concerns or suggestions, it devalues the knowledge and experience of workers who live with the management system, the process, and its equipment. It sets up barriers that can erode a formerly healthy culture. In addition, some individuals in management positions simply do not accept bad news. This can create a significant communication barrier between workers and management and may compound the perception by management that everything in the operation is fine. Workers can lose enthusiasm when they perceive that management does not care.

- How well are you documenting employee concerns and addressing each in some way (even if it is to determine that no action is necessary)?
- How can an organization better provide recognition for those employees who bring solutions that reduce risk or enhance holistic quality in other ways?
- Is there a mechanism for workers to communicate problems and deficiencies in the field without fear of retribution?

3.2.13 A lack of trust in field supervision

When workers feel that they cannot turn to their immediate supervisors with concerns, it indicates a lack of trust. Some of this may be due to a fear of reprisal or embarrassment.

Many issues stop at the level of field supervisor if the facility culture does not embrace a high level of cooperation and trust. An organization that only hears positive news is not hearing the complete, most important news about its operations.

- Is your organization suffering from the results of communications that get lost in the field?
- Does your organization value the leadership and active participation of field supervision, and does the organization equip these positions with leadership training?

3.2.14 Employee opinion surveys give negative feedback

Whether it is a process safety culture specific survey or any number of general employee opinion surveys, negative response suggests a need for organizational follow-up. Recognize that surveys tend to draw out the negative aspects of the workplace since there may not be another outlet.

The worst use of these surveys is to perform them and then not act upon the results. When evaluating such surveys, it is important to look for positive indicators. The fact that some workers react positively to the work culture and others do not could suggest that inconsistent rules exist between areas or different styles of leadership. Carefully analyze workplace surveys in order to draw meaningful conclusions.

- Do employee surveys often result in a large ratio of negative feedback?
- Are clear patterns detected between working groups or departments that indicate localized rather than systemic problems?
- Do employee surveys elicit a low response rate?
- Are there too many employee surveys?
- How do you best determine measurable actions that will address each legitimate employee concern?
- How can the facility take those actions in a way that positively influences employee perception?
- Have opinion surveys provided some positive feedback, and how does this reconcile with the negative feedback? Are such discrepancies associated with certain work groups?

3.2.15 Leadership behavior implies that public reputation is more important than process safety

Sometimes management decisions are intended to influence public perception rather than to address organizational integrity within the operation. Such conflict can negatively affect each employee's level of operational discipline. An obviously insincere philosophy does not inspire the best performers to perform consistently.

- Most organizations have a set of core values and mission statements that define basic operational discipline concepts within their message. Does your leadership act in ways that demonstrate those values and attitudes?

- Is what is said externally to the community consistent with what is said internally to all employees?

3.2.16 Conflicting job priorities

Too many assigned tasks or job duties can result in overlooking critical process safety tasks. Reductions in the workforce have necessarily condensed jobs to include additional tasks. Some workers have dual roles and report to multiple bosses or supervisors. They may often wonder which job is most important. The organization needs to emphasize that process safety is always the priority where it applies and must provide a means of how to resolve the conflicts when they arise.

- Do the production objectives and the safety objectives sometimes seem to be in competition? If so, does production often win?
- Do you feel conflicting priorities confuse employees, and as a result, they overlook critical process safety tasks?
- Do you delegate critical process safety tasks to less experienced employees because the more experienced people are so overloaded?
- Do you identify the process safety related tasks associated with each job position in the facility?
- Does your team effectively train on these tasks (or ensure that raining has taken place)?

3.2.17 Everyone is too busy

When workers are trying to do too much, there is an increased risk that processes and procedures may be overlooked or that errors will occur. It is amazing to consider the impact that each employee can have on resources, both time and money, by causing delays resulting from lack of communication, not understanding the management system, and poor prioritization. The added costs here hide in the fact that day-to-day operations are simply not efficient. Again, this is something sensed by observing the organization. Projects tend to overrun original estimates and tasks are often assigned by co-workers to others who may not be the best qualified for that task. Helping workers understand their roles in the system is a secondary benefit of responding to this warning sign.

- Are all critical tasks being performed as planned on a daily or weekly basis?
- Is there a tendency for workers to pass on their assigned work to others in order to meet deadlines?
- Is the quality of work suffering as a result of workers appearing to be too busy?
- To the degree possible, are all staff members involved in working through the action plans?

3.2.18   Frequent changes in priorities

An organization that frequently alters job priorities may be out of control. In fact, the operation may be controlling the people. Operating in such a reactive mode violates the principles of effective management. This is a significant warning sign.

Facility leadership that regularly alters daily operating plans without good reason is also evidence of this warning sign. Some semblance of normal operations needs to occur in the day-to-day facility setting. In the day-to-day operation of facilities, two forces may seem to be in competition and conflict. Production and quality objectives, representing customer and business needs, may seem to compete with goals for employee and contractor safety. If business goals are not satisfied, we will not be in business for very long; however, if we do not meet safety objectives, businesses will eventually suffer along with our people. Many organizations try to prioritize these objectives by adopting a safety-first mentality. In fact, these seemingly competing objectives are really the same objective from different perspectives. Our customers' needs for goods and services are very dependent on our ability to manage our facilities in a safe manner. In addition, our ability to deliver products and services consistently on time and in specification often creates the stable environment that allows us to achieve a high level of safety performance.

- Is there a formal system for establishing work priorities within each operating area?
- Is the authority of those who establish work processes and standards clear and coordinated?
- Is the system followed?
- When priorities change, are changes communicated to the workforce with the reason for the change?

3.2.19   Conflict between workers and management concerning working conditions

The presence of this warning sign can indicate substandard working conditions. How management responds to these complaints in particular may be one of the most important elements in developing a strong safety culture. Safety based conflicts can represent an engaged workforce and may draw attention to management decisions that may have needed more input from hourly workers. When resolving employee complaints regarding working conditions, it is critically important to involve representatives from the hourly workforce in the process.

- Do you have an engaged joint health and safety committee?
- Does this committee evaluate key procedural or policy-level safe work issues?
- Docs this committee have a direct link to management?

3.2.20   Leaders obviously value activity-based behavior over outcome-based behavior

This warning sign is present when the organization is more focused on documenting the completion of process safety related activities (or any type of business activity) than it is in getting high quality, long lasting results.

- Are leaders participating in work reviews, and are they taking an active interest?
- Does the organization identify measurable activity to plan the work to be done and monitor the evidence necessary to ensure that it has been done?
- How do you actively redirect this behavior in your organization?
- Does management have a role model for thorough implementation of process safety management?

3.2.21 Inappropriate supervisory behavior

If supervisors are exhibiting aggressiveness or obnoxious behavior in directing the workforce, there is a definite need for intervention. A breakdown might have occurred between the two most critical human layers of protection. The human resources department may need to be involved for certain types of redirection.

- Has bullying been observed in the workplace?
- Are certain individuals treated differently than others?
- Has alienation occurred between workers and supervisors?
- Is this problem systemic or is it limited to a small number of specific departments or areas?
- Is there a system for workers to voice their concerns about unfair treatment in the workplace?

3.2.22 Supervisors and leaders not formally prepared for management roles

Often in the process industries, people are promoted to management positions based on their technical strengths and not on their managerial skills. Many times these more subjective managerial skills are lacking. Often, this is the fault of the organization and not the individual. The organization may not have developed in these new leaders the competencies necessary to manage other humans effectively. When an organization forces people into positions of leadership without adequately training them in basic leadership skills, it can result in aggressive or otherwise difficult behaviors as managers. The resulting situation can damage the new manager's career and create a difficult situation for the people they are managing.

- Is a formal screening process in place to ensure that workers considered for promotion are capable of working with people?
- Is a formal succession plan in place for all supervisory personnel to avoid quick response decisions when vacancies develop?
- Does your organization's human resources group use commonly available tools and programs to prepare technical personnel for managerial roles?

- Has the organization provided the training that supervisors and managers need to be successful?
- Does the organization assess managerial proficiency?

3.2.23   A poorly defined chain of command

A chain of command is a prerequisite for communicating instructions in a timely manner. An effective chain of command will ensure that all aspects of a job are assigned and addressed. In emergencies, the chain of command may change depending on circumstances. Leaders should ensure that this is clear at all times.

This warning sign suggests that the organization imperils accurate and timely communication or instruction to all who need it. Essentially, there is an organizational void when people do not know who is in charge. In critical situations, all workers must know where to turn for direction. Decision-making processes are less well defined when you bypass the chain of command, and this can create difficulty in regaining control over an operation. Unless a chain of command is clearly established within a strong team-based organization, confusion will reign. There have been situations where leadership performed the work of the technicians in difficult situations due to a lack of command structure, among other weaknesses.

- Is there an unofficial chain of command reflecting the sentiment—*how things really get done around here*?
- Does upper management at the facility oversee day-to-day operations in a way that encourages and reinforces appropriate communications and approvals?
- Is there a mechanism to communicate changes in the chain of command to the workforce?
- Is the chain of command for emergencies clearly defined?

3.2.24   Workers not aware of or not committed to standards

When a large segment of the workforce cannot tell an auditor the basic purpose of process safety concepts, it is a sign that the culture has not reinforced the tenets of process safety strongly enough. If you observe that basic personnel safety standards such as hearing protection are not managed, this is a sign of a cultural lapse in the organization. Failure to have fundamental standards in place and understood regarding personnel safety, process safety, health and hygiene control represents failure in the basics levels of a safety culture.

- Is there a basic edict within the corporate safety policy that requires employees to follow company standards?
- Are workers informed of new standards?
- Is there a resource available to explain the meaning and relevance of company standards?

3.2.25 Favoritism exists in the organization

Obvious favoritism for individuals or groups can result in bad operating and maintenance decisions. Favoritism shuts down communication with a larger sample of the team and restricts organizational options. It contributes to a general decline in workplace morale.

- Are some workers favored with special assignments that might enhance their job progression?
- Are some workers given more opportunities to work overtime and earn more money than others?
- Have employee opinion surveys indicated that such favoritism exists? If so, did follow-up action address the issue?
- Does the human resources group use current industry-accepted hiring tools to assist in selecting employees for new positions?

3.2.26 A high absenteeism rate

When workers have developed a sense that their presence is optional on any given day, the culture of the organization is often to blame. A specific instance of absenteeism may have a valid reason. However, a general trend indicates worker apathy toward the workplace. This may be due to physical working conditions, the social conditions, or poor hiring practices. High absenteeism may also suggest that workers might be looking for work elsewhere. Finally, high rates of illness suggest discontent in the workplace due to many reasons, including working conditions.

- How do you ensure that new workers are clear on the attendance policies?
- Are facility policies implemented consistently for all workers?

3.2.27 An employee turnover issue exists

There are two things to consider regarding the warning sign of higher-than-normal worker turnover rates:

- Behavioral factors between or within the organization's people, teams, and work groups
- The physical work environment

Turnover rates that are lower than expected are more difficult to define and harder to address. A stagnant workforce may be unwilling to accept new ideas or to progress with time. A stagnant workforce, either management or hourly or both, can be a root cause of another warning sign—normalization of deviance.

- Are worker compensation and benefits at your facility comparable to those at nearby facilities?
- Does your facility evaluate the causes for turnover of employees and implement resulting actions?
- How does the organization keep its staff engaged and involved in learning new roles and responsibilities?

- Have you established a good employee turnover rate for your site?
- Are organizational or personnel changes occurring so often that it is difficult for leaders to get a sense of ownership?

3.2.28 Varying shift team operating practices and protocols

Despite the existence of formal procedures, many activities take place within an organization that are initiated and managed within working groups. Examples include:

- Adherence to visitor sign-in protocols
- Timing of lunch breaks
- Facility walk-through inspection protocols (who, how, when)
- Attendance management
- Operating preferences (such as using two pumps versus one)
- Backup coverage during field activities (such as sampling)
- Lineup of equipment during cleaning
- Control configuration

Some of these activities vary from department to department, or even shift to shift. In fact, team members may be unaware of what happens on alternate shifts. If such deviations are widespread and significant, they can constitute a threat to the operation. The deviations violate the concept of discipline and consistency.

- What is the best way to achieve consistent practices among all work shifts and crews?
- Are the workers aware of the differences between working groups, and do they see this as a problem?

3.2.29 Frequent changes in ownership

When the economic climate is conducive, corporate mergers and acquisitions increase in order to minimize unit production costs. When changes of ownership occur, all workers may be required to adopt a new set of standards and practices. Change in any form represents a challenge to the workforce. Frequent changes in ownership may exacerbate this challenge. A diligent effort to communicate new expectations to the workers can keep them from becoming confused and demotivated. They may be sensing that further changes are on the way. Ultimately, process safety performance can be affected and significant incidents may occur.

- Does your organization or facility have a well-established process safety culture and clearly defined deliverables?
- Is a communication strategy in place to notify the workforce of expectations resulting from changes in ownership?
- Has a management of change system been used to support changes in corporate ownership that may have occurred in your organization?

3.3 CASE STUDY – <u>CHALLENGER</u> SPACE SHUTTLE EXPLOSION IN THE UNITED STATES

In January 1986, the U.S. space shuttle _Challenger_ exploded shortly after liftoff, killing its crew of seven and causing a severe setback to the U.S. space program. The catastrophic incident was the result of a rubber O-ring seal failure between adjacent sections of the solid fuel rocket boosters. Unprecedented cold weather on the day of the launch made the rubber brittle, which, combined with the faulty design of the joint, allowed hot combustion gases from the burning rocket to escape. The flames and hot gases burned through the metal supports holding the rocket in position. When the rocket assembly released, it ruptured the side of the external fuel tanks allowing liquid hydrogen and oxygen to mix prematurely and explode.

During the investigation, it became apparent that there was a well-documented history of problems with the rocket booster design, including the integrity of the rubber O-ring joints. When the NASA team observed O-ring damage after the second shuttle flight, the NASA team simply made changes to the assembly process (but not the design) and continued with future flights. The growing momentum to keep the shuttle flying was adversely affecting the team's reaction to the O-ring problem. NASA middle managers repeatedly violated safety rules requiring the prompt resolution of technical problems. Over time, the managers normalized the deviation, so that it became acceptable and non-deviant to them. Given the success of previous missions with known problems, middle management came to accept the risk and failed to communicate their concerns to top decision makers. While it may have been painful to accept defeat in terms of project schedule, such pain is insignificant when compared to destruction of the mission.

The NASA organization was highly complex. Private consultants were contracted to support important parts of the project, and their livelihood depended on the success of the overall mission. Furthermore, the _Challenger_ project was operated much like a business, with future funding based on past performance. The ability to secure such funding depended on the spreading of good news. Production objectives emphasized the importance of the launch date for each mission. In light of the safety concerns and potential setbacks, work continued on the project. Had the team collectively recommended the suspension of further work until they solved all the technical problems, the catastrophic incident could have been avoided. A strong safety culture would have ensured a diligent and consistent approach to dealing with important issues. The responsibility for making such unpopular decisions in large organizations geared for success is often not clear. Effective leadership should have resulted in someone stepping forward and challenging the status quo without fear of consequence. That is what leadership is all about.

While the _Challenger_ explosion was not a process incident in the strictest sense, the nature of this catastrophic incident was similar to many process safety failures that have occurred in the process industries. In fact, many of the NASA system failures closely align with process safety element failures. The failure of several process safety elements at the same time is symptomatic of a failure in the process safety culture of an organization. Process safety culture is the milepost upon which success in an operation can build. Without it, other initiatives will only be partly effective.

- What actions might have interrupted the chain of events that led to this incident?
- Why was such action not taken?
- Who had the responsibility to act?
- At what point was the incident inevitable?

This incident serves as a classic example of the type of loss that can occur in a large, complex organization if management systems are not effective, and strong team leadership is not promoted.

Exercise: Can you identify warning signs that may have preceded this incident?

4
TRAINING AND COMPETENCY

Learn as if you were to live forever.
Gandhi

4.1 WHAT IS EFFECTIVE TRAINING, AND HOW IS COMPETENCY MEASURED?

Effective job training provides workers with initial and ongoing knowledge and skills to develop the competencies required to perform their jobs safely in a manner that supports quality, environmental responsibility, and economic success. Training is effective only when a change in behavior takes place and competency is assessed using practical demonstration on the job or worksite.

Do the people who operate your facility recognize all of the safety equipment available to them, understand how the equipment functions, and what they have to do to make sure that it is working properly? You should not qualify a person as competent until a competency assessment proves that the person can apply the training. A competency is a collection of knowledge, skills, and attitudes that combine to result in effective performance. Therefore, effective training programs should target these three areas for greatest impact on employee performance.

Competent workers are essential in preventing catastrophic incidents. Both classroom and on-the-job training are essential. Regular verification and application of accurate procedures help confirm that the workforce is, in fact, competent and has the right attitudes, knowledge, and skills to perform its work. Effective training confirms that the workforce applies its training consistently and sees value in it. Management can see the changes by comparing behaviors before and after effective training.

In a process facility operation, a good quality training and skills assessment ultimately enables workers to operate complex systems within the specified safe operating range to meet product specification and, most important, to avoid catastrophic process incidents. Planned formal training will help ensure that workers operate the process facility to its intended design in a safe and efficient manner. Formal training should be designed and conducted to meet the technical design requirements of the operation and the standards of operating stated in the site operating management system.

In many of the case studies provided in this book, the root causes of the events included deficiencies in training and competency. Even more important, as the events began to unfold that led to the catastrophic incident, the workers failed to recognize one or more of the following three things:

- The warning signs that a catastrophic event was imminent
- The speed at which the event occurred
- The potential consequences of the event

Providing training in these three areas is of utmost importance for workers in processes where there is a potential for catastrophic events.

4.1.1   Three basic levels of training

One model for process facility training is made of three parts. A description of each is provided below.

Fundamentals Training: Fundamentals can include topics such as site process hazards, pressure, temperature, flow, general safe work practices, hazard identification and risk assessment, regulatory training, company orientation training, personal protective equipment, emergency response plan procedures, and common processing steps as appropriately indicated by the analysis phase.

Basic Process Training: The process overview training can include topics related to the equipment configuration, process and equipment hazards, chemical and physical changes, and special safe-work practices related to the operations, maintenance, and materials. Emphasis should be given to new equipment and chemical hazards.

Job-specific Training: Job-specific topics include training on new or revised operating, safety, and maintenance procedures. Job-specific training for process safety should include all the systems that control and manage process safety hazards.

A curriculum developed for an operator position at a facility should list the following:

- The fundamental training the employee received upon hiring (or had completed previously)
- Process overview training for each process where personnel are assigned to perform tasks
- Job-specific procedure training for the equipment and safety management systems for each task you expect operators to perform

This curriculum addresses the competencies required on initial assignment to a job and can capture changes within a job assignment as well as training needed when workers cross-train or change assignments.

4.1.2   Competency assessment

Providing suitable competency assessment is essential for workers. Competency assessment is an evaluation of how a worker actually applies the skills and knowledge to identified tasks. People gain skills and knowledge through previous experience, training, and application. It is vital that competency assessments actually check and validate how workers apply these skills in the workplace. Too often, we identify competency by completion of a training course and / or completing a test. This alone does not demonstrate competency. Consider competency assessment a vital part of plant operation, and high quality competency assurance systems include regular, identified competency monitoring.

Compare the following list of example competency assurance systems against your site practices.

- Identification of competencies required for the position's task inventory and any specific roles
- Identification of minimum competency required before appointment
- Process for dealing with workers who fail to pass competency assessment
- Identification of how competencies will be obtained
- Appointment of competency assessors
- Identification of training programs to develop competencies
- Assessment as to whether learning has taken place and competency levels have been achieved (typically assessed by written testing and on-the-job performance testing)
- Refresher competency monitoring and checks
- Training and competency register
- Systematic performance review of training and competency programs

4.2   THE TRAINING AND COMPETENCY RELATED WARNING SIGNS

Warning signs identified with training and competency assessment are listed below.

- No training on possible catastrophic events and their characteristics
- Poor training on hazards of the process operation and the materials involved
- An ineffective or nonexistent formal training program
- Inadequate training on facility chemical processes
- No formal training on process safety systems
- No competency register to indicate the level of competency achieved by each worker

- Inadequate formal training on process-specific equipment operation or maintenance
- Frequent performance errors apparent
- Signs of chaos during process upsets or unusual events
- Workers unfamiliar with facility equipment or procedures
- Frequent process upsets
- Training sessions canceled or postponed
- Procedures performed with a check-the-box mentality

- Long-term workers have not attended recent training
- Training records are not current or are incomplete
- Poor training attendance is tolerated
- Training materials not suitable or instructors not competent
- Inappropriate use or overuse of computer-based training

4.2.1   No training on possible catastrophic events and their characteristics

The types of catastrophic events that could occur at a facility might include:

- Vapor cloud explosion (VCE)
- Fireball
- Boiling liquid expanding vapor explosion (BLEVE)
- Pressure vessel burst (PVB)
- Low temperature embrittlement from auto-refrigeration
- Runaway decomposition
- Runaway reaction
- Loss of cooling
- Release of highly toxic materials
- Auto-ignition

This warning sign will be evident when facility personnel lack basic knowledge of the following items:

- The potential catastrophic events that could occur in their processes
- The systems and controls in place to prevent such events
- The safety critical variables that if exceeded could lead to a catastrophic event
- The warning signs that a catastrophic event is imminent
- The amount of time it takes for the event to occur
- The potential consequences of the event to personnel in the area

Does your organization assess this warning sign's presence through an examination of the content of the training modules and questions to workers on the topics listed above?

4.2.2   Poor training on hazards of the process operation and the materials involved

A key early warning sign is that facility personnel appear to lack adequate understanding of the site hazards, specifically process hazards.

It is vital that facility personnel understand the process safety hazards associated with the site process, the process materials, facility, and equipment hazards. This is especially important for operating and maintenance technicians. Once they begin to understand the hazards, they can better manage the risks by applying and maintaining the controls that manage these risks.

- Does facility leadership ensure that the site training and competency management system captures the process hazards adequately?
- If the training programs are not enough for assurance, do you monitor and check personnel on the job through competency evaluation and assessments?

4.2.3   An ineffective or nonexistent formal training program

This early warning sign indicates that there is no management system in place for training site personnel. This could be indicative of a general lack or application of a management system and a lack of management commitment within the organization.

Validate training with a formal training program for all levels of the site organization, from the leadership team to each hourly position. Identify each position's process safety training needs and the required level of competency. Use a mixture of electronic computer-based training, classroom training, and practical training. This can be tracked and formally documented in a training and competency register.

Evidence of an effective process safety training system would include documents such as training records, qualification records, or the training materials themselves. When there seems to be a mismatch between the training given to workers in similar positions, the curriculum may vary from individual to individual within the same job classification. Operational excellence and overall capability suffer when the leaders do not really know the minimum expectations for their workers' performance levels.

Consider developing a curriculum for each job position in the facility. Its content should reflect where incumbents work and what they do. It is an invaluable tool for an organization concerning work assignment and competency assessment. The other specific topics that should be addressed in the curriculum for each position as it applies to a job position are:

- Process-specific knowledge training
- Process-specific hazard training
- Operating procedures

- Safe work practices
- Maintenance procedures
- Emergency response procedures

If you find evidence of all these things being revised and trained upon every time they change, documentation of training activities is in order. Even doing the minimum training well and completely is a considerable undertaking.

- Is there a formal training program?
- Does it include documentation and verification?
- How does your training program description compare against its implementation?
- Is it solely to meet regulatory requirements?
- Have you analyzed the job tasks to determine the knowledge, skills, and attitudes necessary for all levels of the organization?
- Have you ensured that each level of the organization is receiving the appropriate training?

4.2.4   Inadequate training on facility chemical processes

Although the industry is constantly improving in this area, many facilities still have inadequate training related to the basic operation and design of the equipment their employees operate and maintain. Some sites do not formally train operators on how to read P&IDs even when using them is part of that position's job description. Process overviews often lack the detail or instructional systems design required for helping operators to learn more about the basic physical and chemical processes they are controlling. Workers employed at an industrial site should be trained on the specific mechanics and technology of each process. Years of field experience at another facility will not enable a worker to adequately manage a new operation without fully understanding the how and why of the process. The example below illustrates this need.

- For example, steam reforming of hydrocarbon is critical in the production of ammonia. The process involves the reaction of steam with hydrocarbon at high temperature over catalyst. The operation is mechanically complex and occurs in a large fired heater known as a reformer. Unlike a conventional heater, the parallel tubes in the reformer are filled with catalyst. Each tube in the heater is a sensitive chemical reactor. These tubes' temperature must be monitored and maintained within a safe operating envelope to avoid failure. The steam-to-carbon ratio must also be managed to control the conversion. Since the process control demands of this operation are burdensome, even the most experienced process operator should have additional technical training on the reforming process. Are your operators provided with detailed process overview training on each chemical process at your facility?
- Does your facility have resident experts who can resolve technical problems above and beyond those described in the training materials?

- Do you have separate process overviews for operators, maintenance workers, and contractors, as indicated by the risk-based nature of your processes, so that each group can focus on the process safety requirements unique to those roles and the risks of the process?

4.2.5 No formal training on process safety systems

If no formal training in process safety exists, this could indicate that the site is not managing process safety risks effectively. Process safety training should include all levels of the site organization and should include the following:

- Site process safety hazards
- Management systems in place
- Roles and accountabilities of site personnel in risk assessment
- Hazard assessment (HAZOP, FMEA, and others)
- Different hazards that need to be assessed, such as:
  - o Operating and maintenance task hazards
  - o Worksite hazards
  - o Process safety hazards
  - o Simultaneous operations hazards
- Controls that actually reduce residual risk, such as elimination, substitution, and engineering controls
- Safety critical systems that control process hazards
- Latent conditions that can exist on process systems which can affect the safe operation of the process
- The importance of maintaining operating limits
- Bypassing process alarms and trips using management of change (MOC), risk assessment, and technical approval
- Application of the site control of a work process

It is still possible to meet experienced workers at a facility from operations, maintenance, management, or the technical staff who do not understand the process safety management implications of their job duties. In addition, they may not have a good working knowledge of the site's management systems that support process safety. This can be observed even when these workers participate in critical process safety activities regularly.

An organization benefits from having its employees understand the basic regulations that apply and the internal administrative procedures that customize a facility's implementation methods. Consider the following points:

- Employees are more likely to view process safety as an activity-based as opposed to an outcome-based process.
- Does management support this view by its actions?
- Poor understanding of an employee's role in critical process safety elements such as process hazard analyses, management of change item

reviews, and pre-startup safety reviews is a root cause of warning signs within other elements.

- Why does management often overlook clarifying and supporting competency in process safety roles?

Doing effective training on how management system elements control process safety risks is critical. All people—from site leader to the newest member of the workforce—need to understand their role in process safety. How many of the site employees have documentation of participation in such training?

4.2.6   No competency register to indicate the level of competency achieved by each worker

This early warning sign indicates that there could be no formal training program in place, or that there is no rigor in formally recording training and competency assessment. Identification of the minimum and full competencies required for each site position or role is critical in managing and controlling site process safety. The system of managing and maintaining training and competency registers is a clear demonstration of a formalized process and that training and competency assessment is taking place. Use these registers for internal and external audit and assurance.

- Does your organization use a competency register or skills profile for tracking and monitoring site workers' competency levels when assigning work?

4.2.7   Inadequate formal training on process-specific equipment operation or maintenance

Look for evidence that your facility teaches specific job task procedures to all personnel as indicated by their position, but specifically for the production technicians in the maintenance and operations groups. You may find evidence in the form of materials and records of procedure training, process overview training, or vendor-supplied training courses.

- Does management require the training department periodically to revisit the level of detail in the process hazard training modules for each job position and include additional training as required?
  - o There is significant benefit in training operations employees on the critical operating limits of the equipment in their area and the importance of knowing those limits. This makes decision processes easier during times of abnormal operation.

4.2.8   Frequent performance errors apparent

Frequent errors (including those that cause process trips and upsets) could indicate lack of operator training and competency. If you do not address this warning sign, it could lead to a number of process safety barrier failures. This increases the risks of a catastrophic incident. Training and competency assessment should address the process safety barriers in order for operating personnel to understand them and then to help maintain them.

Validate this by reviewing logbooks, inspection records, and incident reports. Training is not always the problem in performance issues, but it often receives the blame. Also, consider human factors when evaluating this warning sign's existence. A skilled trainer can help management determine when a problem is training related and when it is not.

- Do your site incident investigation recommendations often throw training solutions at issues that are not training related?
- Does your organization perform training needs assessments using an instructional systems design model to avoid providing unnecessary training?
- When training is the correct action, does your facility use an instructional systems design model to achieve it?

Occasional small errors are leading indicators of the potential for more impactful errors. When an organization displays no concern, it supports normalization of deviance.

4.2.9 Signs of chaos during process upsets or unusual events

If you spend enough time in the control rooms of processing facilities such as chemical plants or oil refineries, you will witness the response to a process upset. Each process is unique in the specifics of how to troubleshoot a situation, how to regain process control, and how to know when it is time to initiate shutdown or an emergency shutdown. When there is failure to agree among workers during these situations, conflict may develop. This conflict can compromise making a safe decision. Such conflict may lead to chaos and expose a lack of clear leadership. Observing team communication during a process upset to evaluate its effectiveness in identifying and performing the correct procedures may be useful for evaluating this sign.

- If you observe this warning sign, reevaluate the employee refresher training records.
- Is more frequent training required on facility emergency operating and emergency shutdown procedures?
- Is your facility complexity and risk of failure sufficiently high to justify considering a process simulator as a training tool?
- Does your facility have an abnormal situation management process that includes training to help operations to regain control, go to a safe operating phase, or initiate a safe shutdown?

4.2.10 Workers unfamiliar with facility equipment or procedures

Industrial facilities are typically complex operations where the hazards of the materials and the consequences of deviating from procedures may not be readily apparent. It is incumbent on the employer to assure that employee training is thorough and effective on the hazards of the process before assigning a person to either permanent or temporary work. Conduct an audit of the site training records to determine two things.

- Have training requirements been defined?
- If so, have they been implemented?

Conducting a field audit of employees' knowledge of the process and its hazards can also prove to be effective. Consider the following activities:

- Ask workers to locate a specific procedure. Can they do it easily and quickly?
- Ask workers to identity the steps of an emergency procedure in their area.
- Determine how the workers receive training on revisions to the procedures.
- Is the training more comprehensive when the change is complex?
- Is equipment and piping adequately labeled in the field?
- Evaluate how well the facility trains technicians on how to read and understand P&IDs.
- Observe how outside tasks are coordinated. Did the work process flow naturally?

4.2.11 Frequent process upsets

The warning sign of a process operating near, or outside, its operating limits is a critical finding. Some examples of this are provided in the list below.

- Excessive flaring
- Unstable process conditions
- Relief valves lifting
- Alarms and trips activated
- Poor operating performance

Investigate the causes of instability immediately when they are unrecognized. Continuing to operate a process outside its operating limits increases the potential for process safety incidents.

Excursions that exceed operating limits can be monitored and tracked by site leadership using the site performance management system. A well-designed key performance indicator (KPI) will alert site leadership when the facility is operating consistently outside its limits. This may indicate a training and competency issue with a specific board operator, specific shift, or just poor operating techniques. It may also indicate that an engineering defect exists in the system; if so, a decision as to whether the process is safe to continue to operate until the defect is cleared needs to be evaluated. There could be physical or design issues, but revisiting training materials and procedures may also be indicated.

- Does the organization evaluate process upset trend reports effectively?
- Is it necessary to investigate further for determining the best-recommended actions?
- Are trends used during incident investigations?
- Are there commonalities in the frequent upsets with unresolved issues?
- Are root causes determined and actions completed?

- Does your operation repeatedly find new ways to introduce chaos into the operation? If so, this indicates systemic management system problems.

4.2.12   Training sessions canceled or postponed

Training should be a top priority, and attendance at scheduled training should reflect this. Resist any attempts to reschedule training sessions unless an emergency exists. This warning sign could indicate that there are no systematic methods of monitoring and tracking. Suggested causes are as follows:

- The site leadership has no tracking or monitoring process in place for training attendance.
- The site planning and scheduling process is poor, and the facility operates in a reactive mode.
- The leadership sees little value in training personnel.

When a site cancels or postpones training courses, the perception of personnel could be that training and competency assessment are not as important to management as other drivers are. If this continues, it can lead to general degradation of competency, therefore increasing the potential of process safety risks. Although there may be legitimate reasons to cancel or postpone process safety related training, it is easy to perceive that an organization has incompatible goals when it slights training in deference to other priorities.

- Does your company review the training schedule in comparison to training attendance records to find anomalies?
- Does the organization's site leadership periodically attend the training to open or close the sessions or access the quality?

4.2.13   Procedures performed with a check-the-box mentality

Do not confuse the good practice of using checklist-based procedures for guiding high performance with evidence of this warning sign. The *check-the-box* mentality intended here is when workers display minimal understanding of the process and process hazards. The organization may promote the workers to follow procedures rigorously even if what they are doing is not working or does not make sense. Observe workers in the control room, at their desks, and in the field as they complete tasks.

- Are there sufficient resources and support to respond to worker questions and problems on all shifts?

- Are workers recognized in a positive way for stopping a job when things are not going in accordance with the plan?
- Do workers understand the process risks of the areas where they are assigned?
- Is worker competence just adherence to procedures, or does it go beyond that?
- Is task analysis used to identify the competencies essential for each job position?
- Is there any indication that maintenance tasks or engineering tasks are also an issue that you could solve in a similar way?

4.2.14 Long-term workers have not attended recent training

Look for this warning sign in the facility training records and in interviews with experienced workers. Ask the following questions:

- Have experienced workers been trained on recent changes?
- Have experienced workers failed to attend job-specific training sessions?
- Do experienced workers sometimes skip training because the training is repeating materials or concepts that the workers already know or think they know?

If the answer to any of these questions is yes, there is evidence of an inadequate refresher-training program and a failure of the training management system. There should be an established frequency for minimum refresher training based on the training needs assessment.

- Do you use on-the-job assessments to allow long-term workers to pass materials they know well?
- Can management actively demonstrate their participation in the required training?
- Are experienced workers consulted on the content, frequency, and adequacy of the training program?

4.2.15 Training records are not current or are incomplete

This warning sign is evident when examining training documentation during a formal audit or an informal review. The recommended records should include documentation of the following:

- Receiving training and assessment on the process overview module(s) when applicable
- Receiving training and demonstrating competency for a process safety related job (and training on all subsequent revisions to those procedures)
- Training on changes to the process and how your site assessed competency (Depending on the scope of the change, this may range from signing the training class roll to taking an exam and demonstrating competency in the field.)

- Receiving refresher training and competency assessment on current operating procedures
- Any special hazard refresher training sessions (for example, tabletop exercises using the emergency response procedures for situations that could result in a catastrophic event)
- The employer consulting with a sample of employees on the frequency necessary for refresher training
- That training and competency assessment is being monitored and tracked by site leadership

If any of these records are missing, it is evidence that leaders may not recognize training as an essential element of the site's program to prevent catastrophic events. During an audit, if there is no record of training attendance, then one should assume that the training was not performed.

- Do you require your training department to ensure that the training implementation phase includes strict adherence to training documentation for all training activities?

4.2.16 Poor training attendance is tolerated

There is an easy way to evaluate the existence of this warning sign. If a site has mandatory annual (or periodic) regulatory training requirements, the only acceptable metric for attendance is 100%.

- Look at training records. Did all affected workers attend required training? Is there documentation that participants achieved the learning objectives?
- Check to see if there is any follow-up for nonattendance.
- Is there a system to hold management accountable for deficiencies in the training program?

4.2.17 Training materials not suitable or instructors not competent

The training materials should be continually upgraded and improved. Instructors should have demonstrated communication and training skills. Where these elements are deficient, it is an indicator that management does not recognize that training is the fundamental tool to prevent catastrophic events. Failure to adequately train is also a failure to manage. If the training materials seem sparse, unorganized, not based upon a documented job task analysis, or are based upon outdated process information, they are probably unsuitable for supporting an operational excellence program. If you do not select instructors for their knowledge and competency in the subject matter and then develop their instructor skills, this warning sign is present.

- Is a defendable system in place to select internal trainers? Does it consider both technical knowledge and the communication skills required to be an effective instructor?

- Does your training department use an instructional systems design model to revisit each job position's job task analysis to determine the knowledge, skills, and attitudes that support high-quality performance?
- Does management support revising training as indicated by such an analysis? If not, this warning sign may be evident.

4.2.18 Inappropriate use or overuse of computer-based training

Computer-based training (CBT) is a self-paced learning system accessible by computer or handheld device. The use of CBT is becoming more common especially on small sites with limited training resources or on sites where a large number of workers must be trained in a short period. CBT has its advantages, but it lacks personal contact between instructor and student.

Merely requiring an employee to take a CBT module for a specific training program could range from having them read a simple PowerPoint presentation with one question at the end—*do you understand this?*—to a truly well-designed and well-developed CBT module to test the higher aspects of knowledge needed for performing critical tasks.

Good CBT modules include valid test questions designed in accordance with the training model. They can feed results directly to the site's employee learning management database and actually document a change in the participant's behavior.

Some organizations have overused CBT such that workers have minimal contact with skilled instructors. Regulatory agencies are often skeptical about the use of CBT. This is especially so in situations where traditional training methods have been discontinued. Citations have been issued in some jurisdictions when companies have been unable to verify the effectiveness of training. Training should be a balance of instructor-led training with CBT. If CBT is used, follow it with instructor interface time.

- Have you evaluated the effectiveness of CBT in your operation?
- Are you willing and able to make changes and improvements to CBT when required to do so?
- Is an instructor readily available to help or assist when questions arise during CBT?

4.3 CASE STUDY – GAS PLANT VAPOR CLOUD EXPLOSION IN AUSTRALIA

In September 1998, a massive explosion occurred at the Esso Longford gas plant in southeastern Australia. It killed two workers and injured another eight. A large part of the plant was destroyed, resulting in a gas supply disruption to the State of Victoria that lasted for several weeks. Financial losses exceeded US $1 billion.

The Esso Longford complex consisted of three separate gas plants built and modified over a period of 20 years. Offshore platforms located off the southeast coast of Australia supplied oil and gas. The facility separated the oil from gas and routed the various fractions to product pipelines for distribution to commercial and domestic markets. A central control room was located adjacent to plant 1, the facility that had been constructed in 1969.

The facility was physically large and appeared to contain state-of-the-art technology and equipment. Recent modifications included cold box refrigeration to separate various gas fractions. In the ten years preceding the incident, the operation had been plagued with line blockages and restrictions due to ice buildup. This phenomenon, common in gas plants, is caused by hydrate formation at low temperatures. This is the result of residual moisture condensing and freezing within the piping. Hydrate formation can be controlled by heating or by carefully adjusting flow rates. If hydrate problems are not addressed or controlled, a gas plant operation is highly vulnerable to upset conditions. Surprisingly, during the ten years preceding the incident, hydrate problems and process upsets were frequently encountered but seldom reported or investigated. These became a normal part of the operation.

During a period of cold weather in 1998, the facility encountered a process upset on a fractionation column. This event occurred on a school holiday when many technical staff members were off-site. The shift leader that day was a maintenance supervisor who had limited experience with facility operations. Attempting to repair a leak on a control valve, employees made several process adjustments throughout the morning. They made these in direct response to the symptoms of the problem. Procedures did not exist. Workers did not follow any set plan for this scenario. After four hours of trial-and-error attempts to re-establish circulation, a worker started a pump that introduced hot naphtha into an extremely cold heat exchanger under pressure. The heat exchanger immediately fractured due to low temperature embrittlement. The failure released a large cloud of flammable hydrocarbon. Several seconds later, the vapor ignited and destroyed much of the facility while causing two fatalities.

The catastrophic incident at Longford did not just happen. The owner modified the facility several times to increase capacity. Reductions in the work force count had taken place without adequate assessment of the impact of the organizational change. The following warning signs were revealed after a thorough investigation:

- Facility process hazard analyses (PHAs) were not conducted.
- There were few formal operating procedures.
- A lengthy history of process upsets relating to hydrate pluggage in lines was recorded but not investigated. Had such problems been recognized and addressed, the catastrophic incident might have been avoided.

- Workers and management at the site were not trained in the phenomena of low-temperature embrittlement and on the catastrophic event that could result.
- There was no system to manage organizational changes resulting in assigning temporary supervisory responsibilities.
- The plant supervisor on the day of the incident was not adequately trained in process operations.
- Instruments and alarms were not properly calibrated or commissioned.

In a large, complex facility such as a gas plant, failures and unprecedented events can occur. Operating personnel should be trained to identify the warning signs of a catastrophic event. Workers and supervision must thoroughly understand the properties and hazards of the process materials. Only with a thorough understanding of the hazards will they be able to respond effectively to the warning signs. The site's training program should include catastrophic event precursors, the hazards resulting from latent conditions, warning signs, and consequences. This type of training can only occur after the facility conducts a process hazard analysis and writes accurate operating and maintenance procedures for each critical task. The PHA results and the procedures become the basis for developing and implementing a successful training program.

Exercise: Can you identify warning signs that may have preceded this incident?

5
PROCESS SAFETY INFORMATION

Knowledge is of two kinds. We know a subject ourselves, or we know where we can find information upon it.
Samuel Johnson

5.1 CRITICAL INFORMATION TO IDENTIFY HAZARDS AND MANAGE RISK

Process safety information (PSI) is a system for documenting and communicating information on hazards relating to chemicals, equipment, and technology. In the terminology of Risk Based Process Safety (RBPS), it is *process knowledge*. Process safety information enables the facility to identify and understand hazards. It documents a rational basis for the safe design, operation, maintenance, and retirement of process facilities. All designs, procedures, operating strategies, training programs, maintenance plans, and management decisions need to reflect a knowledge and awareness of process safety information detailing current process hazards in each operating area. Up-to-date documentation regarding process materials, chemical properties and inventories, operating parameters, equipment details, and associated functions should all be easily accessible to each employee who needs it. Process safety information ultimately supports informed decisions and responsible actions.

The following list includes process safety information common to many processes.

- Process description, including materials, and process chemistry, including credible undesired chemical reactions
- Material and energy balances
- Material safety data sheets (MSDS)
- Hazards associated with internal process streams (including toxicity, permissible exposure limits, physical properties, reactivity data, corrosivity data, thermal and chemical stability data, hazardous effects of

inadvertent mixing of materials normally contained in process streams, utility systems, and typical contaminants—such as air and water)

- Range of material inventories (including maximum intended inventories)
- Safe upper and lower limits for process parameters such as temperatures, pressures, flows, or compositions
- Consequences of deviations from safe limits
- Piping and instrument diagrams (P&IDs)
- Block flow diagram or process flow diagrams (PFDs), depending on the complexity of the process
- Electrical area classification diagrams (also referred to as hazardous area classification diagrams in some jurisdictions) and supporting data
- Relief system design basis and calculations, including flare system design basis
- Ventilation system design basis
- Materials of construction for principal hardware
- Equipment data sheets and vendor data
- Unit plot plans
- Safety systems (for example, interlocks, detection, or suppression systems)
- Facility siting analysis documentation
- Programmable logic controller information
- List of energy sources (for example, high-voltage electricity, explosive mixtures)
- Hazard and risk register
- Process line list
- Company-specific design codes and standards used to form the design basis of the plant or process

A company should formally define each type of process safety information required for any facility. A formal policy typically includes a list of the required PSI, the PSI owners (those accountable for its maintenance, accuracy, completeness, and consistency), the location of each PSI document, and auditing methods to ensure compliance with the policy. The policy should address implementation, including use of document management systems and training of PSI users. PSI requirements will depend on the technology, the types of hazards, the complexity of the process and equipment, the dependency on outside resources (for example, mutual aid groups), and local regulations.

5.2 THE PROCESS SAFETY INFORMATION RELATED WARNING SIGNS

Some common process safety information related warning signs are provided below.

- Piping and instrument diagrams do not reflect current field conditions
- Incomplete documentation about safety systems

- Inadequate documentation of chemical hazards
- Low precision and accuracy of process safety information documentation other than piping and instrument diagrams
- Material safety data sheets or equipment data sheets not current
- Process safety information not readily available
- Incomplete electrical area classification drawings
- Poor equipment labeling or tagging
- Inconsistent drawing formats and protocols
- Problems with document control for process safety information
- No formal ownership established for process safety information
- No process alarm management system

5.2.1 Piping and instrument diagrams do not reflect current field conditions

Maintaining an up-to-date set of P&IDs requires each facility to have an assigned work process with the required budget and management direction to maintain the site P&IDs. For some mature facilities, there simply may have been too many unmanaged changes over the years. This can create a problem that the organization might consider as *too huge to tackle right now*. Leaders regularly agree to delay it indefinitely. Older facilities may not have all electronic CAD drawings for the entire site and the condition of some original drawings is so poor that they are not legible or usable.

However, accurate P&IDs are an essential element for even beginning to manage process safety. P&IDs play a critical role in the safe operation of a facility.

- They are used in developing operating procedures.
- They are used for training process operators.
- They are a visual communication vehicle or roadmap for making changes or adjustments to an operation.
- They provide a flow schematic for the process.
- They are a quick source of technical information for control room panel operators and field operators.
- They are used to identify hazards during periodic process hazards analyses (PHAs).

When P&ID accuracy erodes, they may not reflect the equipment in the field in a usable way. Operating decisions using the P&ID may be in error, such as in the following scenario:

1. A field operator contacts the control room about a leaking connection on a heat exchanger.
2. The control room operator instructs the field operator to bypass the heat exchanger using the upstream manifold.

3. However, unknown to the control room operator, an auxiliary bypass was recently installed upstream of the heat exchanger and was not recorded on the P&ID.
4. When the field operator opens the bypass, the process stream is routed to an alternative and unintended location.

P&IDs often indicate equipment labeling and nomenclature as well as technical details about categories of equipment. Rated pump capacities, pressure safety valve set points, and instrument ranges are a few examples. Finally, control loop and shutdown schemes are typically documented, interpreted, and understood using P&IDs.

In some jurisdictions, local safety regulations specifically require that drawings and data sheets accurately reflect equipment in use in the field. When it becomes apparent that some drawings are inaccurate or out of date, the integrity of the entire process safety information system may become suspect. This may trigger scrutiny by regulators. In addition, local facility personnel may lose confidence in the P&IDs as well.

- What is the status of your site's P&IDs? Can they be relied upon? Have incidents occurred because P&IDs could not be used in the field?
- Do you allow errors on P&IDs? How are they corrected?
- Are changes approved by the designated owner of the P&IDs?

Inaccurate P&IDs pose a risk to an operation, but they are also symptomatic of other issues. They are a strong indicator that the management of change system may not be effective. Field changes are made without the necessary hazard assessment, discipline, and follow-up. The operations department can be a verification resource for periodic cross-checking of the field equipment against P&IDs for consistency.

- Do you ensure that P&IDs are maintained current and accurate through a formal management of change process?
- Does your facility have an appropriate document management system in place to maintain accuracy?
- PSI documentation should be valued similar to the asset value. P&IDs are the core of PSI and are the most difficult facility documents to keep current. Reevaluating P&IDs often requires field verifications that consume resources and funds.

5.2.2   Incomplete documentation about safety systems

Safety systems perform a dedicated designed safety function. Some are preventive in nature (e.g., relief valves), while others mitigate the effects of an incident once it has occurred (for example, dikes and fire suppression systems). Safety systems are critical systems. They are required to function with a high degree of reliability. When an organization has let critical information related to safety systems erode, it is a sign that management attention needs redirection.

Safety systems such as relief valves and shutdown systems require a schedule of preventative maintenance to ensure high reliability. Documentation on safety systems helps ensure completion of preventive maintenance. Documentation on safety systems is also important information to have during hazard assessments to verify that adequate layers of protection are in place. Some jurisdictions require accurate documentation on safety systems including the design basis. If the documentation related to safety systems, is inadequate or incomplete, ask yourself the following questions:

- Have you performed a line-item review of safety systems and their minimum process safety information requirements?
- Does the site ensure that documentation of safety systems is managed through a formal management of change process and an appropriate document management system?
- How do you correct deficient data in the PSI system?

5.2.3   Inadequate documentation of chemical hazards

Chemical facilities typically handle, store, and process large quantities of hazardous materials. For good reason, most operating jurisdictions require that companies communicate chemical hazards to workers. Often, material safety data sheets (MSDSs), available for all commercially supplied chemicals, are used for this purpose. They indicate physical and chemical properties as well as hazards and safety precautions to be taken. Inadequate documentation about the chemical hazards that exist at the facility may lead to incidents.

Material safety data sheets themselves do not adequately ensure that workers know the chemical hazards. Workers should know the physical location of hazardous material inventories and their respective quantities. The quantity of material is significant since it can affect the size of an incident and the distance at which harm may occur. Finally, workers should know the potential reactivity between different chemicals. Sometimes, a single chemical substance when heated beyond its normal stability point may react violently causing highly hazardous consequences. Such was the case in the Concept Sciences explosion, which claimed five lives in 1999. This involved violent decomposition of a material when processed at temperatures and concentrations above its safe limits. A strong understanding of chemical hazards is vital during hazard assessments to ensure that adequate layers of protection exist to prevent catastrophic incidents.

Although chemical facilities have formal operating and emergency response procedures, emergencies may require third-party and mutual aid responders to assist in major incident situations. These potential workers unfamiliar with the operation need documentation to help ensure their safety. This documentation must include the location and inventories of hazardous materials.

In any operating facility there are chemical byproducts and waste streams. Sometimes these are aggregated until they can be removed. They may be highly hazardous and are usually not included among supplier material safety data sheets.

Where possible a facility should generate internal MSDS for these materials. Where these materials cannot be identified, measures should be implemented to prevent contact with workers.

- Within the process, there may be intermediate products that accumulate within equipment. These can expose workers to hazards upon opening equipment. Should you test these materials and document the results so that workers can take the necessary precautions?
- Do you include intermediates in the site procedure for PSI requirements and ensure that this subject is addressed during PSI user training?

5.2.4   Low precision and accuracy of process safety information documentation other than piping and instrument diagrams

Inaccurate PSI documentation can lead to field errors by engineering, operatoring and maintenance personnel. In the context of normal operation, an engineer may specify a part based on incorrect information. If maintenance then installs that part and operations uses it, it could fail catastrophically. Inaccurate documentation increases the likelihood of making such errors. Applicable documents can include the following:

- Data sheets
- Process flow diagrams (PFDs)
- Plot plans
- Electrical area classification drawings
- Isometric drawings
- Line designation tables
- Tie-in lists (graphic)
- Shutdown keys (graphic)

This information can be critical for safe operation. Accurate documentation of all types is a requirement for hazard assessment, analyzing emergency situations, and assessing safety, environmental, quality and other compliance issues.

From the maintenance perspective, inaccurate drawings can lead to procurement of incorrect materials. Moreover, such errors may undermine the confidence of workers. They may choose not to use the drawings in the future.

- Do all block flow diagrams, process flow diagrams (PFDs), process equipment drawings, and all training-related drawings reflect the current process configuration?

5.2.5   Material safety data sheets or equipment data sheets not current

Material safety data sheets are used to convey safety information about chemicals, chemical compounds, and chemical mixtures. If these are not up to date or do not reflect current conditions, errors may be made.

Many jurisdictions have regulations that require material safety data sheets for all raw material, intermediates, and finished products at a facility. As conditions change, these will need to be updated or changed. The safety functions of mechanical equipment should also be clearly communicated on equipment data sheets. Safety data sheets often highlight special precautions to be followed when using or maintaining equipment. Material safety data sheets and equipment data sheets are also used during hazard assessments. Finally, data sheets sometimes provide a reference to spare parts. When this critical information is inadequate or missing, substitutions could occur which might lead to failure that can result in a catastrophic incident.

- How do you verify that the hard copy or electronic compilation of material safety data sheets is up to date?
- How do you verify that equipment data sheets are updated for facility changes?
- How do you ensure that this documentation is included in the management of change and the document management systems?

5.2.6   Process safety information not readily available

Process safety information contains information that is critical to hazard assessment and safe operating decisions. Since PSI often originates within the technical and engineering groups, there may be a tendency for it to reside in those locations. However, the primary user of PSI is the operations department. It is used to ensure that they maintain the process within safe operating limits. The operations staff is less likely to use process safety information unless it is readily accessible to them.

Process safety information is often not used on a daily basis but rather ad hoc, as it is required. However, it should be available without delay. Often in the past, process safety information was stored in hard copy form and normally resided in a file cabinet or engineering room adjacent to the control room. More recently, however, many companies have provided electronic access for storing and retrieval of process safety information.

- How accessible is your process safety information? Is it the most current? Is there a risk of older, out-of-date information being used? How do you prevent this?
- Have people been trained on how to access PSI? Do they praise the ease of use?

Do operations and others have access to the most current PSI? How do you know? Are the most current redlines available to them? A PSI protocol can specify the location of each PSI document and address implementation, including use of document management systems and training of PSI users.

5.2.7   Incomplete electrical / hazardous area classification drawings

Electrical / harardous area classification drawings are used to allocate the type of electrical equipment that may be used in a particular section of a facility. Electrical area classification is a risk-based system that examines the fire hazard from a fuel release perspective and designates acceptable electrical hardware based on its ignition spark potential. The electrical area classification scheme for a given facility is represented on a plot plan.

Electrical area classification, referred to as hazardous area classification in some jurisdictions, is an important concept for preventing the introduction of an ignition source into an operating area and for controlling maintenance activities. Determine the electrical area classification system appropriate for your site. Electrical equipment installed in designated hazardous areas must conform to requirements specified on the electrical classification drawings. Electrical work in designated hazardous areas must also conform to the zone or division requirements specified on the drawings. Many facilities issue work permits based on electrical area classification requirements. One final point related to electrical area classification is very important. The calculations upon which the electrical area classification is based should be filed with other process safety documentation. Examination of this documentation is part of many process safety audit protocols.

Incomplete or inaccurate electrical area classification drawings can result in two conditions that increase the probability of catastrophic incidents.

1. The installation or use of electrical equipment that is not properly rated for the area and introduces an ignition source into the operating area
2. Errors in implementing the work permit system that could introduce an ignition risk in a hazardous operating facility

Here are some questions to consider when addressing this warning sign:

- Do operating staff understand the principles of electrical classification? Are they able to identify classified and unclassified equipment?
- Are your electrical area classification drawings managed through a formal management of change process?
- Do you refer to electrical area classification drawings when specifying new electrical equipment?
- Do you use the electrical area classification drawings for risk assessment and control of work specifically in the management and control of hot work?
- Do you audit against your electrical classification drawings to ensure compliance?

5.2.8   Poor equipment labeling or tagging

In a perfect world, every valve and piece of equipment is labeled with its unique identifier in the field. In the real world, this is uncommon. However, workers are more effective when they can double check their procedure driven work against the equipment label.

Equipment labeling is an essential part of any safety program. Unless equipment is labeled clearly and visibly so that it is distinguishable from similar equipment elsewhere, errors may occur. Manufacturers generally supply equipment with a nameplate that provides mechanical design information. This is not sufficient for purposes of identification in the field. Ask the following questions:

- Does the tagging conform to the facility's general equipment numbering system?
- Is it clearly visible from the location where work generally takes place?
- Are all vessels, heat exchangers, pumps, piping, and instrumentation labeled?
- Are electrical switchgear components and breaker panels labeled?
- Is color-coded tape or paint used on piping systems with an arrow designating the normal direction of flow?
- Are colors used to indicate the process service for lines?
- Is there accurate and up-to-date equipment tagging and labeling throughout the site?
- Is it a requirement to reverify the information periodically?
- Is labeling part of new equipment installation?
- Are workers trained on what the different tags and labels mean?
- Do workers create their own system by painting or marking equipment?

One potential consequence of poor equipment labeling is human error. Workers will tend to rely on experience and judgment when conducting field activity if the equipment is not labeled clearly. Even an experienced operator will occasionally work on the wrong equipment if it is not clearly distinguished from surrounding equipment. This problem may be magnified in a large, complex facility with multiple similar process units. Workers involved in a process startup or process upset may also be vulnerable to misidentification of equipment when labels are unclear. Equipment identification is important for tenured operators, new hires, maintenance workers, and contractors. A single piece of mislabeled equipment can lead to a very serious incident.

- Have you evaluated the need for upgrading equipment tagging and identification accuracy and methods?

5.2.9 Inconsistent drawing formats and protocols

The presence of this warning sign indicates that an organization does not have, or has not enforced, its own formatting requirements and may have allowed outside organizations to use their own systems and formats. This often results in inconsistency and varying levels of quality.

- Does your site enforce one set of drawing standards? If not, initiate this immediately for all new drawings.
- Does your facility have a system in place to revise and upgrade each inconsistent drawing over a realistic period or as it is involved in a change?

5.2.10 Problems with document control for process safety information

The primary concept of effective document management consists of these points:

- Ensure that document users have access, at all times, to all approved current documentation applicable to their job tasks.
- Ensure that out-of-date documentation is removed from the system.
- Ensure that people understand and have been trained on PSI document management policies and procedures and understand their responsibilities.

Whether the organization manages compliance with a hard copy document management system or an electronic document management system (EDMS), these points are necessary to help ensure safe operations. It sounds simple, but many good organizations often find that their process safety information document control is less than effective.

Effective distribution of process safety information to the user community ensures that workers will have the latest documentation relating to the process. Distribution is a process that requires checks and balances to ensure that users receive and acknowledge new information. Upon receipt of new drawings and data sheets, remove obsolete information from circulation. Failure to do so could lead to confusion and contribute to errors.

In some facilities, workers hoard redundant process safety information for personal use. Personnel computer files, lockers, and file cabinets often contain documentation that is out-of-date. The master drawing file may contain duplicate copies of the same drawing, albeit different versions. These are merely symptoms of problems with the overall distribution system. When such problems occur, there is usually a reason for them. Before merely correcting the immediate problem, make an effort to determine why the problem has occurred. This might expose further management system issues.

- Has your organization established the most effective document control management system suited for your organization?
- How will it be managed continuously for all process safety related information?
- Have you determined the best ways to make the system easy for users to find specific approved PSI managed within the document control system?

5.2.11 No formal ownership established for process safety information

Management systems function efficiently only if organizations assign specific responsibilities. There must be a clear owner for each process safety element and an organizational commitment from everyone to comply with the PSI policies and procedures. This includes anyone who affects process safety, including maintenance workers performing field changes. A small change such as adding a drain can have large consequences. Process safety information can be a costly endeavor for many operations if it requires the periodic review and updating of thousands of documents. Unless coordinated and maintained well, there will be voids and gaps in the PSI. For a new facility, process safety information is often current and consistent with the physical plant. As changes and modifications take place, a disciplined system is required to ensure the integrity of the PSI documentation. This must include redlining of field changes, process safety element owner approval of these redline changes, and as-built incorporation of the changes in an appropriate time frame. During the interim period until the as-built drawings are completed and ready, the most current redlines must be available for use.

PSI does not keep itself updated. It should be a commitment by all in the organization. If formal ownership for process safety information has not been established at a facility, it is likely that there will be quality and follow-up problems within the system. These, in turn, can lead to a serious incident.

One possible contributor to out-of-date drawings is inadequate liaison and communication between the engineering and operating organizations. All employees must understand the safety importance of current documentation. Updating drawings is an unappealing job for many people unless they fully appreciate and acknowledge its significance to safe operations.

- Are engineers and designers aware of their important role in PSI development and maintenance?
- Do engineering managers place a high priority on updating drawings in a timely manner, or is it considered fill-in work?
- Are engineers and designers sent to the unit to review drawings with those who will be using them regularly?
- Have you performed a PSI element audit to determine whether such quality problems exist?
- How do you ensure that updating PSI is a well-understood requirement of the management of change process?
- Does your PSI protocol define an accountable owner for each PSI document category?

5.2.12 No process alarm management system

The ability of an operator to respond effectively to a process upset or abnormal situations provides an important layer of protection in the prevention of incidents.

The process may be reliant on an operator's response to an alarm signal. The objective of an alarm management process is to help ensure that we control potential hazardous process conditions in the proper priority. Alarm management includes an alarm rationalization process where appropriate alarm rates per hour, alarm priorities, and settings are determined based on the potential consequence severity and required response time by the operator. A cross-functional team with knowledge of the process and operation is generally used to perform this study. Alarm management also includes documenting corrective actions for key alarms as part of standard operating procedures.

The absence of a process alarm management system may contribute to operator confusion or less than ideal performance. This may lead to missing an alarm during process upset conditions and could result in equipment damage or a loss of containment. In addition, the absence of an effective alarm management system can lead to stress during process upset conditions and could contribute to an improper response to the condition by the operator.

- As with all proper industry practices, any deviation from the alarm management process needs to be properly documented and approved at the correct level within the organization.
- Has your site addressed alarm management?
- Are changes to alarm priorities and settings part of a management of change process?
- Have you considered and addressed suppressed alarms, device alarms, bypassed alarms, acceptable rate of alarms during abnormal situations, alarm philosophy, and operating target sheets with strategy to regain control if the operation goes beyond the operating windows?

5.3   CASE STUDY – BATCH STILL FIRE AND EXPLOSION IN THE UK

A fire and explosion occurred at the Hickson and Welch chemical works at Castleford, England, in late 1992. Five workers died, 18 were injured, and significant damage occurred to the process facility and nearby buildings. Some of the casualties and damage occurred on an adjacent plot believed to be a safe distance from the primary process. The facility has since been shut down and dismantled.

The facility had been in operation since 1961 and specialized in the production of aromatic compounds that were highly toxic and flammable. The facility covered an area of 175 acres and employed several hundred workers. A large office complex and employee parking lot were located approximately 400 feet from the main process area.

The incident occurred in a process used to manufacture nitrotoluenes. A batch still consisting of a horizontal drum heated with an internal steam coil heated the raw materials to reaction temperature. Over a period of 30 years, the operation had changed, but byproduct sludge had accumulated to a depth of 14 inches on the floor of the still. Finally, the site decided to clean out the equipment during a maintenance turnaround. Unfortunately, this proved to be a difficult task.

A worker attempted to sample the sludge to determine its consistency. However, there was no evidence that a sample was sent for analysis nor was the atmosphere inside the vessel checked for flammable vapor. It was mistakenly thought that the material was a thermally stable tar. In fact, it contained nitrocresol byproducts that are thermally unstable and highly flammable.

In order to soften the sludge, steam was passed through the coil at the bottom of the still. Advice was given not to exceed 90°C. Workers started the clean-out operation using a metal rake. This makeshift procedure continued for at least one hour. After several workers had taken a break, one worker returned to the job and was burned by an intense flame emitting from the horizontal manway of the vessel. This fire caused several fatalities, injuries, and considerable destruction within a range of a few hundred feet. Several smaller fires contributed further to the destruction.

There were several underlying causes of this incident. Such is the case with most large-scale catastrophic incidents. One noteworthy contributor was the lack of formal chemical data on the residue in the still. Over the course of thirty years, accurate records had not been kept on the process nor were any attempts made to test the materials in the process periodically for hazardous properties. A significant process knowledge void had placed the entire operation at risk. Following the fire, it was not determined whether the sludge byproducts were even stable under normal operating conditions. The hazardous properties of all process materials— raw materials, intermediates, wastes, and products—must be thoroughly tested and published for the use of all employees and contract workers.

While process safety information is a highlight of this case study, other process safety deficiencies were also apparent. A management of change protocol should have been applied before any attempt was made to remove the unknown sludge. The incorrect procedure developed on the worksite contributed directly to the event. Finally, training of workers and compliance to procedures are extremely important even when facilities are out of service.

Exercise: Can you identify warning signs that may have preceded this incident?

6
PROCEDURES

There is always a best way of doing everything.
Ralph Waldo Emerson

6.1   SAFE AND CONSISTENT OPERATION

Procedures are written instructions for performing work to achieve a desired result. Procedures help ensure that all work is performed in a correct, safe, and consistent manner. You can avoid costly errors that could lead to disastrous results by following procedures that specify the steps required to carry out an operation the right way every time. Well-written procedures should include equipment operation details, hazards, and special precautions. The level of detail in procedures will vary based on the complexity of the task and the potential consequences of error. In some cases it may be necessary to provide step-by-step instruction, while in other cases a broad overview may be sufficient. Procedures are especially important in high hazard situations or when dealing with complex equipment. They are also important when several different workers are required to work together to perform common tasks or activities

Procedures provide essential support for process activities in an operating plant. For example, starting up a compressor or commissioning a fired heater requires close adherence to procedures. Procedures should also be used for maintenance, procurement, construction, and equipment testing. Finally, they can be used to support management work processes (such as planning, communications, and process monitoring).

Without written procedures, facilities have no assurance that the organization's intended methods would be used by each worker, or even that an individual worker would consistently execute a particular task in the intended manner each time that he or she does it. Written procedures should:

- Describe the intended activity as well as the process or equipment

- Describe controls in sufficient detail that employees can understand how process hazards are managed
- Provide instructions for troubleshooting when the process does not respond as expected
- Specify when an emergency shutdown should be executed
- Address special situations such as temporary operation with specific equipment out of service
- Describe consequences if certain steps are not followed and when processes deviate from the operating limits, and the steps required to avoid or correct such deviations
- Control activities such as periodic cleaning of process equipment, preparing equipment for maintenance activities, and other routine activities
- Define steps required to safely start up, operate, and shut down processes, including emergency shutdown

The use of procedures in industry comprises a system in itself. The scope includes the identification of need, the writing process, the verification process, the document control process, training, and the ultimate compliance and follow-up in the field. Despite difficulties in developing and maintaining procedures, many process facilities have an abundance of procedures. The biggest problem may reside in the way in which an organization manages compliance with its procedures. Another significant problem area is the way an organization communicates the procedural information and manages procedural compliance.

There is a broad range of procedures (from work practices to checklists) that provides direction and guidance to employees. Some example categories of procedures are provided in the list below.

- Standard operating procedures
- Startup procedures
- Shutdown procedures
- Temporary procedures
- Maintenance procedures
- Emergency procedures
- Safe work practice procedures
- Communication protocols
- Job planning

A number of warning signs derive from incidents with root causes linked to procedural problems.

6.2   THE PROCEDURE-RELATED WARNING SIGNS

The procedure-based warning signs are provided in the list below.

- Procedures do not address all required equipment
- Procedures do not maintain a safe operating envelope
- Operators appear unfamiliar with procedures or how to use them
- A significant number of events, resulting in auto-initiated trips and shutdowns
- No system to gauge whether procedures have been followed
- Facility access procedures not consistently applied or enforced
- Inadequate shift turnover communication
- Poor-quality shift logs
- Failure to follow company procedures is tolerated
- Chronic problems with the work permit system
- Inadequate or poor quality procedures
- No system for determining which activities need written procedures
- No established administrative procedure and style guide for writing and revising procedures

6.2.1   Procedures do not address all required equipment

Procedures are written instructions describing the steps needed to safely operate process equipment. Well-written procedures detail recommended operating practices and safe operating limits of processes and equipment, describe consequences of deviation beyond these operating limits, list actions required to bring equipment back into safe operating mode, and list steps necessary to safely startup and shut down equipment. If critical procedures fail to reference all equipment in the process, operators may not have the information necessary to operate the equipment safely.

Here are some indicators of this warning sign:

- Procedures are not reviewed against current P&IDs.
- Procedures are not audited as part of the facility quality assurance quality control (QA/QC) program.

This warning sign—procedures failing to address all necessary equipment— also leads to another sign discussed later: clear evidence that workers are not using the procedures. The following steps are a test to determine if this warning sign is present. It also helps evaluate the extent of the problem.

1. Obtain a current set of P&IDs for a unit along with a comprehensive startup procedure for a unit.
2. Mark off each piece of equipment on the P&ID as it is addressed by the procedure.
3. When finished, check the P&ID to see if the procedure failed to mention critical equipment. Was anything necessary for startup not marked?

4. Investigate each instance to determine the cause of the discrepancies.
5. Repeat this exercise with other procedures.
6. Analyze your findings to determine procedure program accuracy.

Here are some questions to ask related to this warning sign:

- Does your facility consider stand-alone checklists to be procedures?
- Is there a system to determine which job tasks require procedures?

6.2.2   Procedures do not maintain a safe operating envelope

Safe operating limits are normally set for critical process parameters, such as temperature, pressure, level, or flow. Safe operating limits specify a predetermined boundary for safe operation. The consequences of exceeding a safe operating limit may be catastrophic, such that the risk of continued operation is unacceptable. Workers must understand that any deviation beyond these safe operating limits have potential consequences and risks. These risks and consequences of deviation must be addressed specifically within the procedure.

Procedures should also provide clear and simple instructions to discontinue troubleshooting efforts and take effective and immediate steps to shut down the process or to put the process in a safe operating mode. Whenever operators investigate and take corrective action, written procedures allow them to understand the potential consequences of various courses of action and their combinations. Operators equipped with a clear understanding of the risks involved, and supported by a culture that promotes a healthy sense of vulnerability, are more likely to recognize dangerous situations and take the appropriate actions. Operator understanding is more likely when written procedures include process hazards and when those hazards are highlighted during training. Reemphasizing the hazards at key steps and clearly stating the consequences of deviating outside safe operating limits helps operators understand risk. If the procedures include descriptions of the process control actions, interlocks, and safety systems related to the specific process step, it could enhance performance.

Here are some indicators of this warning sign:

- Procedures do not list safe operating limits.
- Safe operating limits listed in procedures do not match those identified in other process safety information.
- Emergency shutdown procedures have not been developed for critical operations.
- Procedures do not identify *exit points*—the conditions or operating limits that exceed normal operating conditions—from normal operations to emergency operations.
- Procedures do not list the consequences of a deviation and the steps required to correct a deviation or process upset when normal operating limits have been exceeded.

- Procedures do not include a process for reporting deviations, no matter how slight, in a non-punitive way; nor is there a process to follow up and take corrective action on reported deviations.
- Procedures do not specify which operators have authority to shut down a process during an emergency.
- Operators must first seek and gain approval before shutting down the process in a safe and timely manner.

When this warning sign is present, it indicates that the procedures were either not developed with accurate process safety information or that the process safety information changed without adequate management of change.

- Have the site personnel reviewed procedure content versus process safety information to verify that critical equipment is addressed?
- Have the site personnel verified that all parameter targets and ranges within each procedure match the process safety information?

6.2.3 Operators appear unfamiliar with procedures or how to use them

Procedures that people do not follow are of little value. Working from memory or using alternatives or shortcuts to approved procedures can lead to highly unpredictable and unsafe operation. To ensure their use, procedures must be available to the operators at all times, should be maintained so that they are current and accurate, and should provide clear, concise instructions with tasks grouped in a logical manner. In addition, employees must clearly understand and recognize the value and importance of following procedures, not only for their safety but also for the safety of their co-workers and for the mission of the company.

Using procedures to guide hands-on training also helps to ensure that operators become familiar with the procedure as well as how to use the procedure to perform a task correctly. Using procedures as a training aid will help familiarize operators with the content and format of procedures and will allow trainers or other subject matter experts to identify confusing or poorly organized instructions.

Here are some indicators of the warning sign associated with use and familiarity of procedures:

- The written procedure steps do not match actual work practices.
- Employees do not understand the consequences of not following procedures.
- The operator training program does not require mandatory use and review of operating procedures.
- Operators do not participate in the periodic review and update of procedures.
- Procedures are not referenced during supervisor walkarounds or during job observations.

This warning sign leads in several directions. Are the procedures wrong? Have workers been trained on how to find procedures? Have workers been trained on how to use procedures? Do employees understand the value and importance of procedures?

- Do you test this by requesting workers to retrieve the current version of a procedure and observe their ease of meeting the request?
- Do you revisit the job task inventory to determine why a given performance-related situation exists?
- Do you observe employees periodically following a particular procedure and asking the employee to explain it?

6.2.4   A significant number of events resulting in auto-initiated trips and shutdowns

This warning sign could be present due to mechanical, electrical, or instrument failures, but often it is due to a matter of workers not referring to the procedures where limits and techniques could direct smoother operations. Well-written procedures list the safe operating limits of a process, identify consequences of deviations from these limits, and provide instructions for troubleshooting when the process does not respond as expected. Automated trips and shutdowns are initiated to prevent operating outside the established limits. Evidence of a significant number of events that result in activation of trips and shutdown devices may be a sign that procedural weaknesses exist.

Here are some indicators of this warning sign:

- Procedures do not list safe operating limits of the process.
- Procedures do not identify consequences of deviating from operating limits.
- Procedures are inaccurate.
- Procedures are confusing and trigger mistakes.
- Operators are not easily able to access the procedures.
- Facility culture does not encourage procedure use.
- Procedures are not analyzed in relation to tasks that were occurring during the trips or events to determine if modifications or corrections may be required.

A high frequency of these types of trips and shutdowns indicate that there is an issue with maintaining the operating envelope. It might mean that the procedures need to be upgraded. Alternatively, it could mean that workers are not using the procedures.

Ask these questions to determine why the workers might not use the documents:

- Are the procedures inaccurate?
- Are users easily able to access the procedures?

- Is there a culture that does not encourage procedure use?
- Is it necessary to revisit procedures related to tasks that were ongoing during the trips or events? Do you:
  o Modify and approve the documents revisited if required?
  o Train workers on them?
  o Encourage workers to use the procedure while performing the tasks and elicit comments to improve the procedure?
  o Track the trip events in the following six months after taking action to monitor the situation and encourage users to access the procedures?

6.2.5   No system to gauge whether procedures have been followed

A strong process safety culture cannot tolerate shortcuts or any other deviations between written procedures and normal practice. Management should be accountable for establishing a culture of conformance to procedures and standards and cultivating it throughout the entire site team. Employees should be accountable for procedure conformance, not results based on circumstances that they cannot always control, such as output, yield, or whatever it takes to get the product out the door. All employees should clearly understand the high safety and business value placed on procedures and on following them to the same degree as management.

Here are some indicators of this warning sign:

- Procedures are not validated periodically to verify that actual practice conforms to intended practice.
  o Validation often includes field verification by both a subject matter expert and a qualified operator.
  o The procedure should reflect the proper way to complete a task, not merely the way it has always been done.
- Procedures are not available to operators at all times. Otherwise, operators are forced to work from memory and may use alternatives to procedures. Both of these cases can lead to unpredictable, and sometimes unsafe, operation.
- Procedures are not maintained. Procedures should be current to ensure process safety and effective operation. The accuracy and effectiveness of static procedures in a dynamic operating environment will decay rapidly in time and may result in undesired deviations.
- Errors and omissions in procedures are not corrected in a timely manner. Procedures that are well written and at an appropriate level of detail are more likely to be used. Long delays in correcting errors or omissions may send a message that *close enough* is acceptable.

Some sites use the technique of interactive procedures for those tasks critical to safety, environmental compliance, and quality. The term *interactive* indicates that the user is required to have the procedure in-hand while performing it, and some input is required to log performance.

Which procedure tasks require sign-off? This issue can benefit from a risk analysis. Not all procedures need sign-off or timing records, so all the procedures do not need systematic check-off. Generally, startup and shutdown procedures are interactive as well as any other critical task identified by a risk-based analysis. For example, the procedure for starting up a critical compressor may need an interactive procedure that requires sign-offs and submittal for filing, but starting a simpler filter system may not require the procedure to be in-hand and signed off.

- Does your facility use an interactive procedure format when indicated by a task's risk level? If so, consider the following things:
  o Include spaces for recording the start and stop times of subsections, and provide sign-off spaces for a team member to use to track completion of each step.
  o Mark any changes or corrections while performing the task. If a step cannot be completed as written, ensure management approval of the field change to the procedure before proceeding.
  o The completed marked-up procedure should be submitted to the appropriate individual for assessment.
  o Revise the procedure if necessary.

6.2.6   Facility access procedures not consistently applied or enforced

A safe work practice on personnel access typically addresses this issue. Management should establish clear rules governing access to operator work areas and process areas. Anyone desiring entry into a process area should first notify the responsible operator, state their intentions, and get the operator's permission before proceeding. This helps ensure that the entrant has the proper permits and protective equipment and will not interfere with or face hazards from other activities in the area. It also helps ensure their accountability should an emergency occur. Visitors and contractors should also notify the operator when they leave the area. In the event of an emergency, this will help account for everyone and saves emergency responders from searching for people who are not present. To minimize both safety and security risks, operators should maintain surveillance of the area and challenge anyone who is present without permission.

Access to control rooms and other operator and contractor work areas is monitored in a similar manner. The mere presence of unnecessary personnel interferes with the free movement of the operations staff, and the distraction level increases with loud conversations, noises, and irrelevant questions. The visitors may cause operational upsets by using their radios near sensitive electronics or by leaning and sitting on control panels if adequate seating is not available. Here are some indicators of this warning sign:

- Operators are unaware of personnel within their areas.

- No procedure exists that requires operators to sign off on visitors or work activities performed in their areas.

This warning sign means that management has not emphasized one very basic requirement of a process safety management system. The act of not controlling and monitoring personnel in harm's way displays a low level of operational discipline regarding concern for the safety of others. The site needs the ability to log each person's location and condition as soon as possible after an incident.

- Have you revisited the safe work practice on controlled area access requirements to find any governmental or organizational requirements that may have been missed?
- If so, revise the procedure and retrain the affected personnel.
- Monitor performance after the procedure training and reemphasize the importance of adhering to procedures.

6.2.7   Inadequate shift turnover communication

Everyone depends on communication to exchange information, and reliable communication is essential to reliable operation. Accurate, timely, and complete information transmittal during turnover is essential. Establishing formal minimum communication standards between shifts will minimize the potential for misunderstanding. This is particularly important during non-routine activities such as startups, shutdowns, and facility outages. Many incidents have occurred because of a breakdown in communication between shifts.

Here are some indicators of this warning sign:

- No minimum requirements and guidelines for effective shift turnover
- No training for employees in their role as communicators for critical operations activities
- No logbooks established for key operating positions
- No written guidance or standard form for recording information
- No periodic review by shift supervisors for accuracy, completeness, and to emphasize their importance
- Late log entries
- Outgoing shift does not orally brief the incoming shifts

A best practice at some facilities is to develop a protocol for good shift turnover with written minimum requirements. It is used as a training document, and different units can modify their documents as needed. Some facilities also stagger the relief times of the supervisors and operators to allow the oncoming supervisor to be more actively involved in the shift relief of the operating personnel.

- Do you have a safe work practice on minimum requirements and guidelines for effective shift turnover meetings?

- Have you trained everyone in their role as communicators for this critical operations activity?

6.2.8   Poor-quality shift logs

Having employees maintain a written record of operations in a structured logbook is one of the most common and reliable ways to ensure that oncoming workers understand the status of the equipment and work activities when they take control. In addition to the importance of maintaining intershift communication, shift logs are also useful for building a database of performance information.

Here are some indicators of this warning sign:

- Handwritten logs are incomplete and illegible.
- Format and content are inconsistent over time.
- Logs are ambiguous.
- Shift supervisors do not periodically review logs for accuracy and completeness to emphasize their importance.

Modern electronic operations logs are becoming more commonplace, but when logs (or forms) require handwritten entries, legibility is a basic problem. Sometimes the format of the information changes between entries. This makes it harder to understand. Moreover, many companies complain of a low level of detail in the log content.

- Have you addressed this warning sign by developing a safe work practice on minimum requirements and guidelines for using the shift log, whether electronic or paper based?
- Are specific shift turnover instructions available for each process or area in a large facility?
- If you have an electronic shift log, do you analyze it for usability by the staff and management?
- Are all workers trained for their role in this critical operations information-sharing activity?

6.2.9   Failure to follow company procedures is tolerated

If the facility staff holds the perception that it is acceptable to bypass the procedure system, it is a sign that the organization, its teams, and its workers' level of operational discipline can improve. If any shortcut, bypass, or deviation from any procedure is tolerated or unaddressed by any facility employee, it sends a signal that the company is not serious about the value of procedures to process safety or the operation of the business.

A high level of operational discipline is critical to ensuring that procedures are followed. This is a cultural issue. Management must insist that workers follow procedures by encouraging their use and reprimanding those who do not use the proper procedure. Management personnel should ensure that they follow procedures themselves and provide the necessary resources to keep procedures current. All employees need to have the same understanding about the reasons for and value of procedures and not simply comply for fear of penalty.

Here are some indicators of this warning sign:

- Incident reports reflect a high number of incidents where lack of procedural compliance was cited as a contributor.
- Retraining on the use of procedures is not provided.
- Managers and supervisors do not promote the use of approved procedures.
- Workers are not encouraged to offer corrections to inaccurate procedures as they find them during use.
- Supervisors do not get out into the plant to monitor procedure use.
- No formal system is in place to verify that procedures have been followed or why they were not followed.

The work that went into developing the existing procedure system at the site is wasted if the procedures are not used. In addition, whatever level of quality the procedures hold today will only diminish over time if workers do not use and maintain the procedures. It is the organization's responsibility to provide the priority, emphasis, and resources to help workers *want* to have high-quality procedures.

Ask the following questions to address this warning sign:

- Should you consider performing a training needs analysis against the job task inventory to see if retraining is required?
- Do you regularly retrain workers on the use of procedures?
- Do you routinely evaluate the use of procedures and attempt to determine why some procedures may not be followed? Is follow-up initiated?
- How do you get managers and supervisors to begin encouraging use of approved procedure-based systems?
- How do you encourage workers to offer corrections to inaccurate procedures as they find them during use?
- Do supervisors monitor procedure use during facility walkarounds?

6.2.10 Chronic problems with the work permit system

An effective work permitting system is a fundamental component of a good safe work environment. Work permits allow operating organizations to control and coordinate planned physical work activities in operating areas. A work permit is essentially a contract between two or more parties that signifies knowledge of work scope, job hazards, and a commitment to following certain protocols and procedures to control the risks. A work permit is a written document. Work permits stipulate conditions under which work may be performed safely. They are specific to a certain scope of work, specific to the equipment or part of a process to be worked on, and specific to who may perform that work. Complex jobs involving more than one task by different persons may require several permits.

Work permits have a finite life and typically expire at the end of a work shift. Work permits can be terminated following a facility incident or upon a breach of the terms in the permit. The work permit system allows operating personnel to make important decisions with a clear knowledge of what work is taking place and where people are located. Responsible persons should physically handle and exchange the work permit. The permit issuer is responsible for fully understanding the scope of work and for ensuring that the permit considers and reflects actual facility conditions. This means physically visiting the site of the work to verify the safety of the job plan. The permit must be carefully filed and be readily accessible while a job is underway. Finally, at the completion of a job, the permit must be signed and returned to the originator. This indicates that the job has been completed accurately and safely.

In practice, the work permit system can be compromised and shortcuts can develop. When some safety professionals see this, they call it an *armchair permit*. This case can exist in larger facilities where a considerable amount of mechanical work may be taking place. Work permits may default to a lower priority than the actual process operational risk would indicate. Many facilities audit work permit system compliance on a regular basis, especially during turnaround periods and other times of high work-level activity. Failures of the system during these times can have catastrophic consequences.

Managing a large number of permits on a daily basis requires considerable organization and discipline. There has been a tendency in some operations to pre-write the permits in anticipation of numerous contract workers arriving all at once at the control room or permit shelter during the start of the workday. This practice can put the operation at high risk. Rubber-stamped permits cannot satisfy the true intent of the permit process.

Here are some indicators of this warning sign:

- Failure to fully define the scope of the job in the field
- Failure to formally train and test permit issuers and receivers
- Copying previous permits for similar types of jobs
- Pre-signing permits before the applicant has actually requested approval from the permit issuer
- No dialogue between the permit issuer and the worker who will be doing the job
- Permits not filed in a logical manner or sequence for later reference by operating personnel
- Failure to cross-link different work permits in a common work area or on a single unit of equipment
- Failure to direct the worker to the right piece of equipment in the field
- Failure to provide direct supervision over workers
- Failure to abide by the terms of the permit (for example, control of files, hazards, gas testing, and SCBA requirements)
- Too many tasks on a single permit

- Failure to sign-off permits at the end of a job
- Different units have different permit requirements
- Using morning gas-test results for work occurring several hours after the initial checks
- Failure to follow the permit-to-work procedure
- Failure to revoke or reissue a permit when problems arise in the field or when the scope of the job changes
- Failure to visit the job site to ensure its safety

It is good practice to record work permit infractions through frequent audits and assign action items to address them. Many organizations use nonconformance reporting systems, and a high level of discipline is applied to ensure that such problems do not recur.

Under many countries' process safety regulations, the work permit system is also a safe work practice procedure. The safe work practices as a group are both operating procedures and maintenance procedures since they apply to all work in the facility when appropriate. If workers report quality issues such as inconsistency between similar work event permits or lax enforcement of site permit requirements, their reports represent a significant warning sign that risk levels could increase in an uncontrolled manner at an unexpected time.

- How do you evaluate the work permit system needs, revising the document if needed, and retraining if necessary?
- Do you perform regular audits to help ensure strict compliance?

6.2.11 Inadequate or poor quality procedures

This is a significant warning sign that may contribute to a catastrophic incident. A poor quality procedure is one that fails to accurately instruct a worker on what to do and how or when to do it. The weakness may be in the instruction itself or there may be a disconnect between the procedure and the equipment in the field. Finally, the procedure may lack a suitable explanation to ensure that important steps are followed properly.

The quality and format of procedures varies across industry sector. There is no standard protocol that fits all needs. A good procedure will typically highlight the important steps to follow in the correct sequence. Special precautions to be taken and consequences of not doing activities a certain way should also be highlighted. A procedure is normally written based upon the assumption that the worker using it has acquired the necessary skills through training. The actual format and style should consider the work culture, language issues, entry-level qualifications, and the complexity of the operation. One of the best ways to improve the accuracy of procedures and ensure that they are appropriate for those using them is to involve the employees of the area in the procedural development and periodic review.

Operating and maintenance procedures should provide current, accurate, and useful written instructions for normal operations and non-routine or infrequent tasks. Write procedures in sufficient detail that a qualified technician could perform the task consistently and successfully. Just as written operating procedures help ensure consistent operator performance, written management controls help maintain and continuously improve the quality of operating procedures.

Here are some indicators of this warning sign:

- No management controls exist defining the quality, content, and format of operating procedures.
- Typically, many different people write procedures. With no guidance, inconsistency is the only predictable outcome.
- Failure to establish specific standards may make the procedures harder to use because of variability in format, structure, and content.
- Management controls do not clearly define who has the authority to develop, change, review, and approve operating procedures.
- Operating procedures do not address all operating modes.
- Procedures do not provide clear, concise instructions.
- Procedures provide incomplete instructions or instructions that are not in the appropriate order.
- Vague and conflicting procedures exist for responding to an upset condition.

Procedure upgrade projects are sometimes necessary for a facility that has lost control of its procedure program. Workers who have been intimately involved in these types of intensive upgrade projects often choose to keep the procedures current long after an effort of that magnitude.

- Has your operation used the AIChE/CCPS book *Guidelines for Writing Effective Operating and Maintenance Procedures* to evaluate your current practices?
- Have you conducted a procedure upgrade project based on an updated plant process hazard analysis?

6.2.12 No system for determining which activities need written procedures

Since procedures exist to help improve worker performance, a list of the tasks that people do (or are supposed to do) at the facility is a good starting point for identifying what activities may need formal procedures. When determining if you need a written procedure, consider what happens in the absence of the procedure. Will the task not be performed or be performed incorrectly and inconsistently? If the consequence of not performing a task or performing a task in an arbitrary manner is acceptable, a written procedure may not be required.

The number of procedures required and the level of detail needed in the procedures is typically a function of the risk associated with the activities and the competency of the employees assigned to perform the work.

Failure to provide the necessary procedures will lead to lower-than-desired human reliability. However, too many procedures, or procedures that contain extraneous information or too much detail, are difficult to use. Non-routine operating modes warrant particular attention because they often involve much greater risk than do routine operations. Balance these factors when identifying tasks for which procedures need to be written.

- Processes or activities perceived to be a high risk or a severe hazard even if you consider the risk of an incident to be low usually indicate the need for a written procedure.
- Facilities with an evolving or undetermined process safety culture may require procedures with a higher level of detail to manage risk and help ensure good operating discipline.

This warning sign is often tied to the tendency of leaders at a facility to take upon a—*we know what they do out there*—attitude about identifying procedure titles and training topics.

When handed a complete list of the tasks resulting from a job task analysis, managers are often surprised at the hundreds of tasks that an effective operator knows and performs.

- Do you use an approach similar to a job task analysis to identify all of the tasks that one performed at each job position?
- Do you ask the users to rank the tasks concerning criticality to safe operation, frequency of performance, and difficulty or complexity (either mental or physical)?
- Have you established a site standard for identifying tasks that only need training and those that should be proceduralized?
- Do you investigate all incidents and near misses to examine whether a procedure would have made a difference?

6.2.13 No established administrative procedure and style guide for writing and revising procedures

Many sites do not have an administrative procedure that addresses the regulatory aspects of the procedure programs or a customized style guide for procedure writers and revisers to use to solve quality and consistency issues. This can lead to deterioration of the procedure program quality over time.

Procedures can vary from precise step-by-step instructions that require initialing and data insertion by an operator, to procedures used only for reference when performing routine operations, such as starting or stopping a pump. Procedures should define safe operating limits and include the consequences of deviations from these limits. They should clearly detail the steps required to execute a task. They should not be ambiguous but clear, concise, and in a format that is easily used.

The number of different formats and styles used at a facility is a trade-off. Multiple formats provide procedure writers with the flexibility to use the most appropriate one, but too many different formats and styles can be confusing to procedure users. In addition to specifying the format and style, facilities should also consider the following:

- Specify the minimum content that must be included in the procedures.
- Ensure that all regulatory requirements concerning procedure content are specifically addressed.

Operating procedures for situations that occur infrequently, such as upset or emergency situations, are important since operators will have less experience in dealing with these situations. Failure to establish specific standards may make procedures harder to use because of high variability in format, structure, and content, and could possibly leave an operator without proper written instructions for responding to a process upset or other unsafe condition.

Here are some indicators of this warning sign:

- No format or style guide exists for the writers and reviewers to use when developing or revising procedures.
- An individual or group unfamiliar with facility format and style standards writes the procedures.
- There is no process to review procedures to ensure consistency in format and style.

Because operating procedures are normally developed at the unit level, and personnel often transfer between units, developing a facility-wide standard for operating procedures will help reduce the learning curve for an operator, engineer, or any other worker who transfers from one operating unit to another. If the format, style, and content are consistent across all units, a newly assigned operator is more likely to be able to locate information in operating procedures quickly and efficiently.

- Has your site developed an administrative, programmatic procedure that addresses all process safety related procedures and how to manage and implement them for compliance at your facility?
  o Operating procedures
  o Safe work practices
  o Maintenance procedures
  o Emergency response procedures
- Have you developed a style guide for the writers and reviewers to use when developing or revising procedures?

6.3 CASE STUDY – NUCLEAR PLANT MELTDOWN AND EXPLOSION IN THE UKRAINE

A nuclear core meltdown and hydrogen explosion at a nuclear reactor in a large power plant occurred in April 1986 at Chernobyl in the Ukraine. Thirty fatalities occurred within a few days of the accident, and a plume of radioactive fallout covered much of northern Europe. Authorities recorded as many as 10,000 related deaths (mostly due to cancer) over the next several years. The facility was eventually shut down and abandoned. Over 300,000 people were permanently relocated to a safer region. The Chernobyl accident is the most serious nuclear incident to date.

The facility contained four separate units with a combined capacity of 3000 megawatts. Each reactor contained thousands of enriched uranium pellets enclosed in tubes within a large mass of graphite moderator. In normal operation, the decomposition of uranium produced heat. This generated high-pressure steam. The steam from each reactor powered two large turbine generators. Unlike other nuclear technologies, there was no secondary containment around the reactor or the steam system. This made the operation critically sensitive to procedures. The formal operating procedures did stress the importance of reliable electric power, especially during startup and shutdown.

The incident occurred as one of the reactors was about to shut down for planned maintenance. Despite the fact that the design was very unstable at low capacity, it was decided to test whether the unit could be safely shut down in a decoupled state. This involved running the reactor (pumps and instruments) using the power generated by the turbine. When a process upset occurred, operators attempted to switch to emergency shutdown mode. When this failed, an unexpected and more extreme spike in power output occurred, which led to a reactor vessel rupture and a series of hydrogen explosions. This released several tons of nuclear fuel and fission products. Over 100,000 people evacuated from the surrounding area.

The Chernobyl power plant had been in operation for two years without the capability to ride through the first 60 seconds of a total loss of electric power, an important safety feature. This may have explained the reason for the test, but given the fatalities and destruction, it was not possible to verify this. While there were several contributors to this incident, such as design, MOC, and lack of leadership, the event is primarily attributed to a blatant disregard for formal operating procedures. Few operating procedures existed, and those that did were not normally followed. If existing procedures had been followed, the deviation might not have taken place. This cannot be verified. In the early stages of the upset, operating procedures should have been available to get the unit back under control. Furthermore, emergency procedures should have been executed when a disaster was imminent.

Operating procedures only work when they are tried and tested. If they are not routinely used, workers will have a difficult time following them when suddenly required to do so. The discipline around operating procedures is an important part of ensuring their success. Each operation or facility needs to adopt a system that will work based on the experience and background of the workers.

When workers discard operating procedures, there may be other reasons. If the procedures are not clearly written or are not practical for field use, they can introduce confusion. Workers tend to rely on practices that have worked in the past or that have evolved over time. Finally, procedures need to be supported by training and be kept up to date.

In any operation, conditions may occur that are outside the scope of existing procedures. When this happens, there should be a clear indication of change. Are workers prepared to react to such abnormal situations and take responsible actions? Has emergency response training been provided for such situations? Workers need to hold one another accountable for following standard operating procedures.

Exercise: Can you identify warning signs that may have preceded this incident?

7
ASSET INTEGRITY

The purpose creates the machine.
Arthur Young

7.1 SYSTEMATIC IMPLEMENTATION

Asset integrity and equipment reliability are the systematic implementation of activities ranging from design to operation and maintenance. These activities help ensure that process equipment will be safe and reliable throughout its service life. The primary objective of asset integrity is the reliable performance of equipment designed to safely contain, prevent, or mitigate the consequences of a release of hazardous materials or energy. This means that equipment should not only function as intended, but also that measures must be in place to prevent or protect against loss of containment. The associated warning signs all have the commonality of not supporting the designed level of integrity necessary for the process.

The integrity of a process and its associated equipment will determine the likelihood of a serious loss of containment incident. Process equipment must be designed to withstand normal and upset conditions during the complete operating cycle as well as occasional outages to conduct maintenance. The design should also account for progressive wear and tear due to aging or continuous exposure to process materials at severe operating conditions. The design of physical hardware should recognize critical failure modes. The integrity of all process equipment should be managed through periodic testing, inspection, analysis, and timely maintenance.

The key components of an effective asset integrity management system should include the following attributes:

- Design, fabricate, and install all facilities and equipment in accordance with industry codes and recognized practices.

- Operate facilities and equipment within design tolerances and within the safe operating envelope.
- Routinely inspect and maintain equipment in accordance with industry codes and recognized practices, including manufacturer recommendations where appropriate.
- Analyze equipment failures to determine their cause.
- Conduct all related tasks using trained and qualified individuals who use approved procedures and complete the tasks as scheduled.
- Use high-quality parts and materials, including a system for positive material identification.
- Maintain an equipment archive with up-to-date repair history.
- Safely dismantle and dispose of the facility at the end of its life cycle.

Preventive maintenance (as used in this chapter, this term includes the broad range of inspection, testing, calibration, lubrication, and diagnostic activities necessary to maintain asset integrity) is currently the norm across industry. This affects both the timing and scope of maintenance activities. Previous life-cycle history can be used to determine the frequency of inspection and the work actually done for each category of equipment. If your facilities are attempting to operate with increased run cycles, it may be necessary to do more work during planned inspections. Unplanned outages caused by surprise failures can be very costly and create safety and environmental impacts.

When necessary, facilities can allocate maintenance on a risk-based *need* as determined by the likelihood and consequence of future failure. Risk-based maintenance is more than just a term. It is based on defendable technology and criteria. Risk-based maintenance is often a subset of a predictive-based maintenance program. Predictive-based maintenance strategy measures the condition of equipment in order to assess whether it will fail during some future period and then taking appropriate action to avoid the consequences of the failure.

With regard to predicting equipment failures necessary for the above-referenced risk-based maintenance approach, be cautious. Equipment failure statistics are available for several categories of equipment (including mechanical subcomponents). These statistics are typically compiled from raw failure data submitted to industry technical agencies on a voluntary basis. The data does not reflect the cause of failure or the service the equipment was in at the time of failure. Nonetheless, published failure data give a relative indication of service life and/or failure frequency under normal conditions. There is sometimes a tendency to discount industry failure data as being too pessimistic and conservative. The suggestion that—*we are better than the industry norm*—should not be allowed to drive major operating decisions. When equipment is approaching or has passed its useful service life, special precautions may be needed. These might include additional monitoring and/or contingency plans. In the absence of such measures, failure data can serve as an important warning sign of an incident.

7.2   THE ASSET INTEGRITY RELATED WARNING SIGNS

The following list itemizes the equipment integrity related warning signs. Although several warning signs may seem similar, subtle differences exist. These differences are important and call for the signs to be discussed separately. These warning signs are grouped with similar signs together in the list below.

- Operation continues when safeguards are known to be impaired
- Overdue equipment inspections
- Relief valve testing overdue
- No formal maintenance program
- A run-to-failure philosophy exists
- Maintenance deferred until next budget cycle
- Preventive maintenance activities reduced to save money
- Broken or defective equipment not tagged and still in service
- Multiple and repetitive mechanical failures
- Corrosion and equipment deterioration evident
- A high frequency of leaks
- Installed equipment and hardware do not meet good engineering practices
- Improper application of equipment and hardware allowed
- Facility firewater used to cool process equipment
- Alarm and instrument management not adequately addressed
- Bypassed alarms and safety systems
- Process is operating with out-of-service safety instrumented systems and no risk assessment or management of change
- Critical safety systems not functioning properly or not tested
- Nuisance alarms and trips
- Inadequate practices for establishing equipment criticality
- Working on equipment that is in service
- Temporary or substandard repairs are prevalent
- Inconsistent preventive maintenance implementation
- Equipment repair records not up to date
- Chronic problems with the maintenance planning system
- No formal process to manage equipment deficiencies
- Maintenance jobs not adequately closed out

7.2.1 Operation continues when safeguards are known to be impaired

Safeguards are devices that provide an additional barrier in the event of a functional or mechanical failure. In order to provide adequate protection the availability of the safeguard must be at least, or better than, that of the system it is protecting. The reason for this is simple; the precise timing of a failure is uncertain. When a safeguard is impaired, there is no protection against failure. Operating in such a manner is reckless and irresponsible. There are also legal implications in many jurisdictions, especially if a failure causes death or serious injury.

Always maintain equipment safeguards in a safe and operable state. They may be required at any time.

- Are your facility's safeguards inspected at regular intervals?
- Is there a system in place to deal with safeguards that may be defective or impaired?
- Is there a rule that requires management to be notified when safeguards are impaired?
- Is the operating crew at your facility authorized to shut down an operation when a safeguard is defective or impaired?
- Is there a process to ensure that regulatory requirements for safeguards are met?

7.2.2 Overdue equipment inspections

Maintaining reliable equipment requires that a facility inspect all physical equipment according to a planned schedule based on applicable codes and practices. This is especially important for pressure-coded equipment, which often falls under the jurisdiction of local regulations. Included in this category are pressure vessels, process piping, and pressure relief valves. When you detect material defects or extraordinary wear, repair or replace the equipment. This responsibility resides with all organizations.

Inspections themselves do not prevent incidents. They provide baseline data upon which to make risk-based production decisions. They also trigger the necessary follow-up (repair or replacement) to ensure that a failure is not imminent. Base your operating decisions that influence run length, product quality, or consequence severity on defendable inputs to avoid incidents. Regular inspections are important for all categories of equipment.

Overdue inspections of pressure vessels, relief valves, and other safety devices or critical equipment is a significant concern in the process industries. No facility or integrity program is perfect. A single piece of equipment may be overdue by a few days or weeks due to extenuating circumstances. If this becomes the norm instead of the exception, however, then this warning sign is present. Similarly, a vessel that is months overdue or multiple pieces of equipment being weeks overdue also indicates this warning sign being present. Another aspect of this warning sign is equipment inspections that are overdue because the inspections or tests are rescheduled for a turnaround that keeps getting pushed out, especially when there is no technical basis to demonstrate that the equipment is still safe to operate.

View the situations above with a sense of concern and alarm. If these major equipment categories are in default, then one must also question the integrity of process piping. After all, piping system failures are responsible for a significant number of loss of containment incidents. Piping systems run throughout most facilities, and they can be difficult to inspect.

Some facilities may defend an inspection backlog as a byproduct of the work planning process. However, overdue inspections represent an unknown risk.

- Is there a risk-based justification for operating a facility beyond a normal run cycle?
- Do you use team input to make decisions to continue operating, or do individuals decide this unilaterally?
- Does the facility have a history of not shutting down to conduct scheduled inspections?
- Does the facility seem to have an excessive number of overdue inspections?

7.2.3 Relief valve testing overdue

The existence of relief valves or other relief devices that have passed their required test date is a particularly worrisome situation. Relief valves often are the primary protection against catastrophic overpressure. Because such valves are active protection measures, it is critical that their reliability be verified via testing. This is especially the case where the devices may be in service or there is an environment that could affect their performance. Some facilities have installed dual relief valves on critical process systems so that they can perform the tests without interrupting operations.

Determine what your jurisdiction's codes and requirements are for relief valve testing and establish a method to ensure that you plan and complete these in a timely fashion. Here are some questions to consider:

- If overdue relief valve tests are found, do you consider operating the process with additional safeguards or other precautions until the earliest possible time when the relief device can be tested and certified?
- Do you implement an unscheduled process shutdown as soon as practical to address the overdue device?
- Did you conduct a risk analysis to allow continued operation until the inspection is completed?

7.2.4   No formal maintenance program

If your facility slips into a reactive maintenance mode, this is a warning sign that the risk of a catastrophic incident has increased. A reactive maintenance plan can be summed up as—fix it when it breaks or develop a plan when the need arises. Corrective maintenance (CM) is the common term for reacting to mechanical and control problems after they appear. Without an established protocol, there is a high potential for making errors. Preventive maintenance (PM) includes scheduled tests, inspections, parts replacement, lubrication, and other tasks based upon time, equipment usage hours, or process throughput. Reliability-centered maintenance (RCM) goes a step further and uses extensive industry data and a subset of itself—predictive maintenance (PdM)—to further reduce the likelihood of a breakdown and at the same time, avoid unnecessary maintenance. Predictive maintenance adds specific condition monitoring on equipment to determine best when to do maintenance.

For smaller facilities with limited maintenance and engineering personnel, implementation of predictive, reliability-centered, risk-based, or other types of maintenance approaches to asset reliability is a challenge. To get started the plant can identify its worst performing equipment and discuss issues with the equipment manufacturers or vendors and initiate contact with other facilities that have similar equipment installed. Sometimes reliability improvements are simple but not easily recognized. You will find that most issues have been encountered and previously solved. Discussing issues with other companies and vendors can be helpful. Here are some questions to consider:

- Is your maintenance philosophy based upon the knowledge and experience of the industry?
- Is your maintenance planning system defendable to a third-party review?
- Is your maintenance planning system well documented such that it is not dependent on a few experienced individuals?
- Does your maintenance planning system include the allocation of skilled craftspersons from within your company or from outside contractors?

7.2.5   A run-to-failure philosophy exists

Running process or safety equipment until it fails is a sure warning sign that safety and reliability are not the facility's priorities. Intentionally allowing equipment to break can create unsafe operating conditions that can then lead to catastrophic incidents. There may be several reasons why a run-to-failure philosophy may exist, such as expected facility closing, pending sale, or major reduction or increase in production demand. These reasons may be valid from a business perspective but are questionable from a safety perspective.

For companies that employ a risk-based approach to asset reliability, there will be some equipment where run to failure will make sense. However, this is only acceptable when you have analyzed the consequences of the failure and know that it is acceptable for both safety and production needs. Here are some questions to consider:

- Are workers aware of the potential consequences of failure if equipment is operated to failure?
- Is documentation of the risk analysis supporting this philosophy available to both operations and maintenance personnel?

7.2.6   Maintenance deferred until next budget cycle

Maintenance is often deferred when budgets are constrained. This typically happens toward the end of a calendar year or, in some cases, when a new maintenance manager is appointed who wants to make an immediate positive impact through cost savings. Unfortunately, the risk of failure does not pay attention to a calendar or attempts at a budget-based career boost. Deferral of maintenance for monetary gain or convenience is shortsighted and suggests that leaders may not fully understand risk. Does this happen often at your facility? If this is the case, then we need to look at which maintenance tasks are involved. We also need to question whether the risk of failure in the short term has been fully considered. Here are some questions to consider:

- When this happens, could you address it by establishing a management team sponsored review of the entire situation and application of a risk-ranking tool?
- Does your system require all decisions to postpone maintenance once deemed necessary to be evaluated thoroughly?
- Have you used MOC to evaluate the consequence of the delay?

7.2.7   Preventive maintenance activities reduced to save money

Simple preventive maintenance tasks being dropped or performed less frequently than recommended to increase profit (or reduce costs) is a *pay me now or pay me later* situation. Every facility has had periods of uncertainty in its economic performance. However, cutting back on critical maintenance tasks that are specifically intended to maximize productivity through reliability greatly reduces the reliability of the equipment subject to those tasks. When that equipment is needed for safety purposes, this becomes a significant issue because equipment may not function when needed. Here are some questions to consider:

- Have you created a process for ensuring that the maintenance management team will be able to prevent accounting-related reductions in facility reliability factors?
- Does your system require a risk analysis for all decisions to postpone maintenance?

7.2.8   Broken or defective equipment not tagged and still in service

Broken or defective equipment may contribute to further failure or failures elsewhere. Any operation that does not readily address equipment deficiencies is highly vulnerable to a loss. Whether the deficiency is mechanical or functional, the first sign of failure is itself a warning sign that related failures may soon follow. A pinhole leak on a line seldom occurs in isolation of other problems. There may be other imminent failures in close proximity. In regulated jurisdictions, equipment deficiencies receive a significant number of audit citations. They are often very tangible and hard to conceal. There may be legal implications associated with operating equipment that is deemed unsafe or unfit.

Sometimes circumstances dictate that broken or defective equipment be in service for a very short period, but this should be supported by a risk assessment, additional precautions or safeguards, and a plan to repair or replace the equipment as soon as practicable. A history of this sort of activity, however, indicates that this warning sign is present. Similarly, allowing the equipment to remain in service for longer durations also is evidence of this sign. When broken or defective equipment is left in service without supporting risk assessments, precautions, or definitive plans for repair, the criticality of this warning sign increases dramatically.

Apart from the physical aspects of equipment failure, one should question the process safety culture of any organization that tolerates defective equipment in service. This is essentially turning a blind eye toward obvious risk and is a strong indicator that normalization of deviance is present. Here are some questions to consider:

- When equipment failures are encountered during normal operation, are these recorded in the shift log and communicated to maintenance?
- Do routine walk-through inspections attempt to document defective equipment?
- Are equipment defects promptly referred to maintenance through the work order system?

7.2.9   Multiple and repetitive mechanical failures

Multiple and repetitive mechanical failures that occur with no obvious consequences suggest nothing more than a pattern of good luck. Eventually, a major failure is likely to occur, with serious consequences. Mechanical failures, however small, are incidents in themselves. Do not ignore them. Failing to recognize and address these failures is more than complacency. It may be a form of negligence.

Repetitive failures on a single unit or category of equipment suggest that the wrong equipment may have been selected for the service or that the life cycle is inappropriate. Both of these causes are engineering related and should be thoroughly analyzed. When a recurring failure pattern becomes obvious, it warrants a thorough technical analysis of the failure mechanisms. Not to conduct this root cause analysis and implement effective corrective measures could lead directly to an incident. Repeated expected failures may not only place personnel in harm's way but may affect facility performance and personnel attitudes. The consequences of operating a process with multiple and repeated mechanical failures lead to normalization of deviance and loss of a sense of vulnerability, which could result in acceptance of substandard maintenance and engineering design and practices. This type of culture opens the door for incidents to occur with serious consequences.

- Would repetitive failures indicate poor design or improper application of the design?
- Could multiple common failures be an indication that the equipment installation and operating parameters may not be compatible?

7.2.10   Corrosion and equipment deterioration evident

This warning sign is present when there are obvious leaks, potential leaks, rusting components, unpainted structural features showing wear, and similar conditions prevalent. Have you ever walked into an industrial process facility and immediately felt unsafe? If so, you were likely sensing this warning sign's presence. When a person first encounters this situation, his or her gut instinct signals that there is a need for a high level of awareness. However, if the normalization of deviance factor plays a role, this condition can become invisible to a person working there every day.

- Do you implement the maintenance management system in a rigorous fashion?
- Can the organization ensure that turnarounds address the issues that cause this warning sign's presence?

7.2.11   A high frequency of leaks

A very visible warning sign is a high frequency of leak and spill related incidents. These indicate that the site has not met the primary goal of handling hazardous materials: to keep the materials in the pipes. Containment has failed. When this occurs repeatedly, it may indicate a culture accustomed to these types of incidents. Acceptance of frequent leaks and spills, even if just steam or condensate, can indicate that safe operation of the process is a very low priority. It could also indicate underlying design issues and incorrect materials of construction. These are most likely precursors to a serious incident. The consequences of continuing to operate with frequent leaks and spills may not be recognized until a serious incident happens.

Safety focused companies conduct incident investigations for leaks and spills and implement corrective steps to prevent recurrences. If there are no follow-up or minimal corrective actions for leaks and spills, then asset integrity improvement is not likely to happen. However, the results of positive corrective actions can be measurable in reduced leaks and spills.

- Have the engineering and maintenance groups developed a strategy for addressing leaks in process units?
- Is there extensive use of leak repair kits or temporary patches?

7.2.12   Installed equipment and hardware do not meet good engineering practices

This warning sign is present when the engineering design, whether purposefully or mistakenly, selects under-designed or otherwise improper materials or components for installation. There are many incidents tied to cases where facilities installed a component without paying attention to its material compatibility with existing piping or systems. Such a simple mistake in design specifications or material handling and control can be disastrous. It may be years before the metallurgical incompatibility reveals itself.

In some situations, it may not be readily apparent if substandard equipment is present. Some indicators that this warning sign might be present are provided below.

- Facility engineers and technical personnel are not knowledgeable regarding current codes and standards (or the applicable codes and standards in effect when the process was designed).
- Facility personnel are not readily able to provide the codes and standards and design specifications for process equipment.
- Instead of a consistent type or manufacturer of a specific component being used, a wide variety of component manufacturers are represented (for example, every pump appears to be different, or valves in identical service performing the same function vary).
- Components from low-end manufacturers are used consistently.
- There is excessive emphasis on cost as a selection or design criteria.

The indicators above do not necessarily mean that the warning sign is present, but if noted, further investigation may be warranted.

Substandard equipment in a facility can often be a matter of cost-cutting efforts. Economic pressures on organizations can drive good people to make bad choices and then defend them based on short-term profitability. This warning sign might be present if your project engineers are excessively cost and schedule motivated or if they often use a phrase such as *what are the minimum code or regulatory requirements for...* or complain that the company's engineering standards are excessive. If your site exhibits evidence of this warning sign, consider the following:

- Has your organization developed a process for a thorough design review plus a design review phase of process hazard analysis?
- Do these reviews include verifying that appropriate codes and standards were used?
- Does your site's piping and vessel inspection program include material of construction confirmation and nondestructive testing to confirm material specifications?

7.2.13 Improper application of equipment and hardware allowed

If a facility is using any equipment for an unintended purpose, there is increased risk of it not performing in the manner expected. This can lead to a failure or even a long-term safety issue. For example, in electrically classified process areas, electrical component installation should be closely monitored. Electrical equipment not intended for classified areas can start appearing in classified areas when this warning sign is present. Process equipment that had been retired from a previous service is sometimes installed and re-commissioned in a new application. When this happens, a thorough analysis is required to ensure that the old equipment is suitable for the new service.

- Is a management process in place to review all projects involving reuse or reapplication of equipment outside its original design intent?
- Do you prohibit this organizational behavior in the facility's internal design codes and practices?

7.2.14 Facility firewater used to cool process equipment

Facility firewater is a critical safety system that, when used, can prevent the spread of fire through a process or facility. Firewater is typically an on-demand system. This means that it must be available instantly when required to respond to an emergency. To achieve such high reliability, a firewater system must be dedicated to its purpose. That is, you cannot use it for any purposes other than firefighting.

In some situations, facilities have used firewater to provide supplemental cooling on the external surfaces of vessels and heat exchangers. This is more common during periods of hot weather when cooling systems are unable to meet duty requirements. The diversion of firewater to non-emergency use puts a facility at risk. It is unlikely that the system could also meet emergency fire load requirements in addition to the process cooling demand. Industry should not treat this issue lightly. In addition, the use of firewater for external cooling can lead to external corrosion, shortening the safe operating life of the equipment.

The misuse of firewater in a chemical facility is a sign that someone in authority is unaware of conditions in the field or does not understand the fire risk. Knowledge of such risk should drive responsible risk controls. Firewater misuse is also a sign of low operational discipline. Here are some questions to consider:

- Was a management of change procedure used to initiate firewater cooling in the field? If so, what provisions were considered for handling a real emergency?
- Was the facility designed to run at full capacity during extreme weather conditions? If not, has engineering been notified and requested to address the problem?
- Are emergency systems inspected routinely to ensure that they are available and ready to operate when required?

7.2.15 Alarm and instrument management not adequately addressed

Alarms and instruments form a vital communication link between important parts of the process and the operator. Without properly functioning alarms and instruments, it is difficult to know the operating status of the process and safety equipment. Instruments may be directly linked to automated control functions, or they may indicate to an operator that action needs to be taken. Some instruments are critical to process safety and others are not. If an instrument is in error or fails to communicate an accurate signal, a process upset could result. Ultimately, this could cause a loss of containment incident. An alarm transmits specific information indicating that an operator needs to act. Some alarms complement control functions and shutdown signals.

Instruments and alarms require unique care and attention compared to other categories of equipment. They must be carefully installed and routinely inspected, tested, and calibrated. Unless you do this on a regularly planned basis, the instrument or alarm may not function reliably. Instrument and alarm maintenance and integrity are best handled as a separate program and should be supported by trained technicians. Any attempt to defer this activity until problems arise is unwise and costly in terms of incidents. The failure to manage instruments and alarms properly at a facility is often chronic and is typically identified by examining instrument repair records. Relative to many other warning signs, the neglect of instruments and alarms presents a high risk to the operation.

Additionally, instruments and alarms should be installed and located in the field in a manner that ensures accurate and reliable operations.

If a program specific to maintaining instruments and alarms is not present at your facility, if maintenance and calibration of such equipment is sporadic or as needed, or if a system to ensure that field installation is reviewed for suitability is not present, then this warning sign exists at your facility.

7.2.16 Bypassed alarms and safety systems

When personnel can disable alarms and safety systems without authorization, this warning sign is present. Even if an authorization system is in place, disabling of such systems should be done only when necessary and only for short durations (for example, the time needed to conduct a test or implement a repair). If authorized disabling is a common occurrence or if alarms and system are disabled for long durations, this sign is present. These kinds of activities indicate that someone decided to ignore a set point identified as critical to safety. It could possibly be due to the decision and action of one person. That behavior in itself points us back to organizational culture as a root cause. Sites should have a system to document and authorize bypassing these systems. Here are some questions to consider:

- Is there a clear edict prohibiting the bypassing of alarms and safety systems?

- Do you require formal authorization for bypassing and disabling of alarms and safety systems?
- Does your site have a process to document each of these situations with justification for the risk level accepted and alternate layers of safety provided during the outage?

7.2.17 Process is operating with out-of-service safety instrumented systems and no risk assessment or management of change

Safety instrumented systems (SIS) have been relatively new additions to the overall protection layers of many operating processes. The SIS is an independent level of instrumentation control protection for some processes that pose high risk upon failure. SIS are independent of the basic process control system and are normally designed to meet higher performance requirements. Testing and calibration of the entire SIS functional system from input devices through processors to output devices are necessary to ensure their performance capability. These systems are not used for normal process control of the operation since they are dedicated for safety and equipment protection. These systems are normally off limits to all except specially trained employees who monitor and maintain them. A malfunction of an SIS would indicate that a critical protection layer of the process is inactive. Processes normally would not operate without the SIS in active operation. This warning sign would be present if the operation was continuing to operate with the SIS in maintenance mode or an inactive condition and without the benefit of management of change (MOC) or risk analysis to approve continued operation.

- How does your operation respond if a SIS is found to be inactive— shut down automatically or continue to operate? Is MOC followed and a risk analysis conducted when needed?
- How well are SIS understood by management and workers in your facility, and are they given priority attention?
- Are the reasons for segregation of process control and safety systems understood?

7.2.18 Critical safety systems not functioning properly or not tested

Based upon the history of recent catastrophic incidents, if a critical safety system cannot meet its testing requirements, a facility either needs to shut down the process or alter production levels to meet an engineered measure of safety. When you cannot certify an automated safety system as operable, there is no other responsible choice. When critical safety systems are not functioning or not tested on schedule while the process is operating under normal conditions without a supporting risk assessment, this warning sign is present. Here are some questions to consider:

- Are critical safety system tests considered a high priority? Is the operation prepared to accept a production penalty to achieve compliance?

- Can a simple customized logic tree help staff determine their next steps? Those steps might be special maintenance tasks or process shutdown.

7.2.19 Nuisance alarms and trips

Nuisance alarms (sometimes called spurious alarms) are a significant early warning sign. This situation could indicate that there are issues with maintenance on safety critical equipment, or it can indicate that set levels on instruments have drifted from the original set point. Finally, the original alarm set points might have been in error.

- Spurious alarms can create bad habits from control panel operators, as they get so used to spurious alarms that they can end up accepting alarms too readily, therefore missing an important alarm indication. It can also lead to a habit of bypassing alarms in order to silence constant spurious alarms.
- Some jurisdictions have sent out a guidance document on alarms and trips due to the connection to a number of process safety incidents. This is the guidance:
    o That spurious alarms and alarm flooding are an early warning sign that instruments and control systems may be in need of maintenance or have inherent design issues.
    o Spurious alarms and alarm flooding create a situation where control panel operators develop an acceptance of alarms and the bypassing of them.
    o All alarms need to be investigated by a skilled person.
    o Operators need to demonstrate that they have rigorous processes in managing process plant alarms and the bypassing of them.
    o All bypassing of alarms requires a management of change process that is reviewed and authorized by a technical authority.

Control of alarms and trips is a key indicator of how site process safety management systems are governing this type of process safety risk.

- What is your site's approach to managing nuisance alarms?

7.2.20 Inadequate practices for establishing equipment criticality

Companies attempting to comply with applicable regulations or attempting to apply risk based process safety management often establish equipment or process criticality rankings to help them focus on the more important areas first. Criticality ranking is essential for applying a risk-based approach to any process safety program. However, if the equipment or areas are ranked without rational justification or reference to a process hazard analysis, it can lead to an unseen or unrecognized level of risk.

- Has your plant's criticality ranking system considered risks to personnel, environment, production interruption, equipment damage, and company reputation? Have you reviewed the ranking results?
- When a question over criticality ranking exists, is there a mechanism to challenge it to ensure that high risks are addressed?

7.2.21 Working on equipment that is in service

If there is evidence that the facility regularly bypasses lockout/tagout requirements, safe work permit requirements, or hot work requirements, this warning sign is present. Some companies misuse a deviation form or variance process of one kind or another to allow things that are strictly against normal operational logic, safety standards, and possibly process design. Many incident reports exist where hot taps or other hot work on live equipment resulted in extensive capital loss and fatalities.

Incident and near miss records that show repeated events involving work on equipment that is in service or that involve shortcuts of safe work permit requirements indicate that this warning sign is present. Here are some questions to consider:

- How do you ensure that the staff is applying their safe work practices rigorously?
- Is every request to deviate from a safe work permit thoroughly understood in terms of planning, design, implementation, and all associated hazards and risk?
- Are there standards for positive isolation, double block and bleed, and other lineup requirements? Are these requirements followed?

7.2.22 Temporary or substandard repairs are prevalent

When quick fixes become the norm at a facility, this warning sign exists. An example would be a welded-on or bolted-on leak box on a corroded pipe. Another would be a temporary piping jumper installed to forestall a problem, but it remains in place and operating for much longer than anticipated. Again, if this is the case, it is inviting the risk level to climb to unrealized heights. Here are some questions to consider:

- Can the management of change system be used more robustly whenever a temporary repair change is recognized?
- If these substandard or temporary repairs are bypassing the management of change system, does your facility have a different, much larger scale problem?
- Are temporary repairs well documented, and is an effort made to schedule permanent repairs at the earliest opportunity?

Most processing facilities schedule outages at regular planned intervals to conduct formally planned maintenance activities. These activities may include equipment inspection, restoration, repair, or replacement. Occasionally, a surprise failure may occur outside the shutdown window. While this may force an emergency shutdown, this is not always the case. Temporary repairs are sometimes made to bridge the time gap between the event and the planned shutdown (also called a turnaround) window.

Temporary repairs may put the operation at risk. The word *temporary* suggests short term only, and it may involve shortcuts or quality compromises. Temporary repairs should be carefully analyzed before proceeding. They may require special precautions and frequent monitoring. If the temporary repair presents a higher than normal risk to the operation, additional safeguards may need to be implemented. While one or two temporary repairs may be justified in the context just described, prevalent temporary repairs could signal a potential major incident.

Every operating facility should have guidelines for implementing temporary repairs. These should consider both the risk of failure and other options to fix the problem. Otherwise, temporary repairs will become the rationale for extended run cycles.

7.2.23   Inconsistent preventive maintenance implementation

Regularly scheduled simple preventive maintenance tasks (such as lubrication or filter change) are usually simple operations for skilled workers. However, when you start seeing these tasks' priorities lowered, it indicates that attention to asset integrity is declining. Many process facilities sometimes find themselves focusing on the most critical maintenance necessary to maintain operation, but this can be shortsighted. The basic lubrication and filter servicing tasks that are delayed are often a cause of the fire that is demanding to be put out by pulling people off routine preventive maintenance tasks. Here are some questions to consider:

- Is a standard scope of work defined for conducting preventive maintenance (PM) on each category of equipment?
- Can delaying or dropping a specific preventive maintenance task be justified with a risk analysis?
- Does your preventive maintenance system require changes to address this warning sign?

7.2.24   Equipment repair records not up to date

Even when the maintenance management system is doing what it is supposed to do, a breakdown at the documentation level can wipe away the good physical behaviors of the workforce with this warning sign. Knowing the status of the process and physical plant is essential to effective process safety. Documenting it on paper or electronically is a critical part of knowing the equipment and process as a whole. No records or incomplete records will limit your ability to justify changes in maintenance intervals and make improvements. It will also limit the ability to troubleshoot future equipment problems effectively. Here are some questions to consider:

- Do you use process safety audits to identify where documentation strengths and weaknesses lie?
- If so, is management willing to implement the corrective actions found by the audit team?

7.2.25 Chronic problems with the maintenance planning system

If there are repeated issues with the maintenance planning system, it indicates that something is awry with management's understanding of a critical tool for maintaining process safety levels. When critical tests, inspections, or preventive maintenance activities are missed or scheduled improperly, it can create numerous concerns. It may even affect the insurability of the facility. For major projects that tie into the maintenance planning system, it is critical that every activity that can leverage the opportunity presented by the project work is used to advantage.

- Are allocated priorities on maintenance work orders sometimes overlooked or altered without a valid reason?
- If chronic problems are detected, do you consult with internal and external experts on the system to determine the proper steps?
  - o   Is it the user?
  - o   Is it the program?
  - o   Is it the data entered?
- How do you determine what is causing the issues?

7.2.26 No formal process to manage equipment deficiencies

Equipment deficiencies and issues can arise at any time and often at the most inconvenient time. These deficiencies are normally first observed by facility operating personnel. If there is not a formal process in place to communicate and manage the deficiencies, they may not receive the repair attention they demand. Establishment of and training on a process for managing deficiencies are crucial to maintaining a safe, reliable process. The formal process should include the following steps:

1. Identify and communicate the deficiency.
2. Record, prioritize, and schedule the repair.
3. Document that the repair was completed.
4. Provide feedback to the operation team of the completed repair.

Without a formal process to address deficiencies that all employees can participate in, you can jeopardize timely and effective deficiency repair. A formal process is highly recommended.

- Does the repair process for identified deficiencies work effectively at your facility?
- What is done to improve the process for repair of equipment deficiencies?

7.2.27 Maintenance jobs not adequately closed out

Have you ever encountered a recent maintenance work zone where a grating or plate had not been replaced? You may see that a few bolts are still missing or insulation is not yet reinstalled. If the job has already been closed out, it should make one wonder: *What else was not done on that job?*

- Does your facility use follow-up quality checks on the maintenance performance teams?
  - o Performing a positive maintenance task should not result in a negative safety situation.

7.3 CASE STUDY – REFINERY NAPHTHA FIRE IN THE UNITED STATES

The Tosco Oil Company operated a 140,000-BPD oil refinery at Avon, California, for many years. During the late 1990s, several serious safety incidents occurred and contributed to the eventual transfer of ownership of the facility. One of these catastrophic incidents involved a flash fire on an elevated scaffold while maintenance work was underway to repair a leaking process line connected to the crude column. This fire resulted in four fatalities, one serious injury, and considerable property damage.

In early 1999, a pinhole leak was discovered in the 6-inch naphtha draw off line from a vertical fractionator column above the 100-foot level. Instead of shutting down the facility, an attempt was made to plug the line with the unit in service. When this failed, it was decided to block in the line by closing valves at both ends of the leaking section. Closure of the valves did not alleviate the leak, however. After several days, it was finally recognized that the entire line was in a severely corroded state and would have to be replaced.

Again, with the unit in service, a plan was drawn up to remove the line in sections by using a cold saw. Several work permits were issued over a two-week period. After one large section of piping had been successfully removed a surge of naphtha sprayed several workers on a platform. The naphtha ignited immediately causing a flash fire that engulfed the workers and the upper section of the column.

This serious incident was clearly the result of unsafe work practices. However, the cause of the leak and corrosion within the line was a change to the processing scheme that had altered conditions within the line at least one year earlier. The condition of the line should have been monitored through good maintenance practices before the leak occurred. Furthermore, a safe work plan should have required that the unit be shut down and purged prior to undertaking major mechanical work. There were several serious errors made during the maintenance planning and field execution. As stated previously, maintenance is an important part of ensuring the mechanical integrity of an operation. When you cannot assure the condition of equipment or make repairs in a safe manner, the process or piece of equipment must be shut down.

The two-week time span, which elapsed from the time the first permit was issued until the final incident, is of particular interest. The work did not proceed in the field initially because neither the problem nor the scope of work was adequately defined. As further attempts were made to isolate the line without success, the work strategy became more aggressive. Despite warnings from employees that the work was unsafe, the message did not reach responsible persons who might have aborted the plan in favor of shutting down the facility. The formal investigation cites management pressure as a contributing factor to the incident. Had management visited the site, they might have recognized the problem and made a safe decision for the benefit of all.

- Was management fully informed of the hazards and risks associated with this job?
- Were they prepared to shut down the facility in the interest of safety?
- Did management in any way convey the impression that bad news was unacceptable?
- Does management at your facility take a personal interest in high-risk maintenance jobs?

Exercise: Can you identify warning signs that may have preceded this incident?

8
ANALYZING RISK AND MANAGING CHANGE

It is not the strongest of the species that survive, nor the most intelligent,
but the one most responsive to change.
Charles Darwin

8.1 RISK MANAGEMENT

In an operating facility, risk management is a continuous effort. During the design and construction phase of the facility, specific risks were taken into account and systems were used to manage them. Thereafter, on a day-to-day operating basis, each item that you change, whether in a large way or a small way, presents a new opportunity for your facility to use good risk management practices in applying process safety management techniques.

8.1.1 Hazard identification and risk analysis

The safe operation and maintenance of facilities that manufacture, store, or otherwise use hazardous chemicals requires an effective system to identify hazards and to determine whether the risks associated with the hazards are adequately controlled. The reality is that no facility is 100% free of hazards and their associated risks. An oil refinery is a good example of this. Regardless of the methods used for processing and storing gasoline, a flammable material, it retains the capability to burn and will always present a potential fire risk if hazard management systems become ineffective.

8.1.2 The definitions of hazard and risk

An understanding of the risks posed by a hazard is essential establishing robust systems to manage hazards and their risk. To do this we first need a basic understanding of hazard identification and risk analysis. The definitions of hazard and risk follow.

- *Hazard:* A hazard is an inherent characteristic of a chemical, a physical condition, or an activity that has the potential for causing harm to people, property, or the environment.
- *Risk:* Risk is defined as the combination of the likelihood and the consequence of a specified hazard being realized.

Hazards are all around us. The challenge is to recognize hazards and then to control them. There is a level of risk associated with every hazard.

A hazard is either present or not present. To expand on the gasoline example, a typical above ground storage tank filled to the high-level alarm point with flammable material presents the following hazards (among others):

- Flammability
- Explosivity
- Toxicity
- Environmental damage
- Falling
- Drowning

The storage tank full of gasoline is not a hazard in itself. The characteristics the tank presents are the hazards.

The probability of actually experiencing an incident involving one of the hazards listed above at a given level of magnitude can be reduced (that is, the hazard's risk can be reduced) by several things, such as:

- Good engineering design
- A comprehensive and effective mechanical integrity program
- A set of accurate operating procedures addressing the tasks associated with the tank
- An effective set of safe work practices (including one that defines restricted areas)
- A vigilant environmental monitoring program
- The culture and resources to implement the points above in all behaviors, related to the tank

However, all of the hazards still exist. The risk severity can be reduced by addressing either the probability or the severity of consequences, but it can only be eliminated when the hazard is completely removed. Hazard identification and risk analysis work together to identify the following:

- Hazards that exist
- Consequences that might occur as a result of the hazard
- Likelihood that events might take place that would cause a catastrophic incident with such a consequence
- Identification and implementation of management systems and engineering controls designed to reduce or eliminate the consequence

An efficient and systematic risk analysis, preceded by a thorough hazard identification effort, can increase an organization's ability to manage risk.

Hazard identification pinpoints hazards and any weaknesses in the design and operation of facilities that could lead to major incidents. This process provides organizations with information to help them evaluate their hazards and manage risks. Hazard evaluations can occasionally be performed by a single person when the specific need of the analysis and the perceived hazard of the situation identified are appropriate. The most common and simple but time-consuming way is to conduct a walkthrough or checklist inspection of the workplace. Look at each task and each operation to see which hazards are present.

To improve the effectiveness of identifying hazards, more structured approaches are used. Many tools are available to conduct such evaluations. A process hazard analysis (PHA) is one of the more common and effective tools used within the process industries. Most formal hazard evaluations, such as a PHA, require the combined efforts of an experienced multidisciplinary team. The hazard evaluation team uses the combined experience and judgment of its members along with available data to determine whether the problems identified are serious enough to warrant further actions or controls.

Developing this understanding of risk requires addressing three specific questions concerning hazards associated with a process or operation.

- What could go wrong?
- How bad could it be?
- How often might it happen?

Based upon the level of understanding of answers to these three questions, we can determine how best to manage the hazards in order to minimize risk. As industry has gained experience with the production, use, and handling of hazardous materials, it has documented this experience in the form of management systems such as design standards, procedures, and programs. The foundation for these systems is based on an organization's ability to:

- Identify hazards
- Assess risk
- Develop and implement administrative and engineering management systems and controls throughout the life of a facility

8.1.3   Management of change

Managing change over the life of a facility is one of the key management systems industry has adopted to identify hazards and manage risk. If you make an un-reviewed change in a hazardous process, or in a management system designed to control a hazard, the risk of an incident could increase significantly. When changes occur in an operation that is hazardous or contains hazards of any sort, it is imperative that a change process exists to understand, control, and communicate these hazards. The introduction of even the slightest changes can have an undesirable effect on one or more of the management systems that are in place.

Management of change (MOC) requires us to use a level of hazard identification and risk analysis consistent with the scope and complexity of the change. Examine every change to a process for impact on the basic process control systems, operating and maintenance procedures, training modules, and all management systems.

Significant changes in processes are recognized as major and complex. This generally leads to a thorough management of change review. However, seemingly minor changes to processes can sometimes be overlooked as an item that requires an MOC review, or may go unnoticed altogether. As such, you may inadvertently introduce minor but significant compromises to management systems. This emphasizes that the triggers for a management of change process must be comprehensive and clear.

The main goal of an MOC program is to ensure that properly reviewed and authorized change requests identify and ensure the implementation of management systems and controls appropriate to the proposed change. This must include communicating the results to affected personnel. In order for an MOC program to address all potentially significant change situations, the potential changes should first be identified. Once identified, you can evaluate them to determine whether the change:

- Introduces unforeseen new hazards
- Increases the risk associated with a known hazard
- Weakens or eliminates an existing management system

Much like changes to a process, changes or deviations in management systems, whether known or not, can lead to hazards being inadvertently introduced and risk being increased. Therefore, an effective MOC program should address two key areas of focus.

- Hazards associated with new or different processes, designs, equipment, or procedures
- Hazards associated with adherence to (or deviation from) established management systems

8.1.4 What is your role in risk management?

An effective hazard identification and risk assessment program should motivate workers to accept personal responsibility and manage their own safety. Workers should be empowered to stop any situations or behaviors that present hazards to people, the environment, and equipment. Persons initiating change should also ask themselves—*Do I really need to make this change? Does the value of the proposed change justify the risk associated with the increased hazard?*

8.2   THE RISK ANALYSIS AND MANAGEMENT OF CHANGE RELATED WARNING SIGNS

These are the warning signs associated with management of change and risk assessment.

- Weak process hazard analysis practices
- Out-of-service emergency standby systems
- Poor process hazard analysis action item follow-up
- Management of change system used only for major changes
- Backlog of incomplete management of change items
- Excessive delay in closing management of change action items to completion
- Organizational changes not subjected to management of change
- Frequent changes or disruptions in operating plan
- Risk assessments conducted to support decisions already made
- A sense that *we always do it this way*
- Management unwilling to consider change
- Management of change item review and approval lack structure and rigor
- Failure to recognize operational deviations and initiate management of change
- Original facility design used for current modifications
- Temporary changes made permanent without management of change
- Operating creep exists
- Process hazard analysis revalidations are not performed or are inadequate
- Instruments bypassed without adequate management of change
- Little or no corporate guidance on acceptable risk ranking methods
- Risk registry is poorly prepared, nonexistent, or unavailable
- No baseline risk profile for a facility
- Security protocols not enforced consistently

8.2.1   Weak process hazard analysis practices

Process hazard analyses (PHAs) should be performed using an established process that identifies key personnel, necessary documentation, methodology, follow-up actions, and closeout. Without these controls, the analysis could be incomplete. This may result in new hazards not being identified or known hazards being assumed controlled, static, and unchanging. Workers can lose their healthy sense of vulnerability. The result could be a major incident. Here are some indicators of this warning sign:

- People are unsure of their roles and responsibilities in PHA.
- The PHA facilitator is ineffective due to lack of training or appropriate background.
- Process hazard analyses are poorly planned, conducted in the wrong environment, conducted with the wrong people, and are documented incompletely or undocumented.
- Process hazard analyses are always rushed with participants feeling that there was inadequate time for analysis.
- Focus seems to be on meeting a deadline for completion rather than conducting a thorough review.
- Participants leave the PHA feeling uncertain about the results and not feeling that they could participate enough.
- Safeguards are assumed to work and are listed in the PHA without discussion and evaluation by the team.
- Safeguards that are not applicable to the specific hazard scenario are listed regardless of applicability.
- Management is not supportive and does not see value in the PHA process.
- Action items are not clearly assigned to a responsible person with a due date and are not always closed out.
- Unrealistic time constraints are imposed on the PHA teams leading to quick assumptions or invalid conclusions.
- Leadership excludes the action items from the final PHA report.
- A revalidation consists of reviewing and accepting the previous PHA, even if there were multiple changes or incidents during the intervening period.

Here are some questions to consider:

- Have these warning sign indicators been audited?
- How does the site's leadership team address the issues, and how effectively do they address the issues?
- What is the perception of the facility staff concerning the quality and value of PHA practices?

8.2.2 Out-of-service emergency standby systems

Emergency standby systems are there to protect the facility in abnormal situations. You are operating without a key layer of protection without these systems in an active mode. The risk to the facility increases and the potential for a catastrophic event increases. Remove these systems from service using management of change and audit the status regularly with a written report to senior management. These systems include, but are not limited to, items such as suppressed alarms, use of simulated values, and safety system bypasses. Here are some indicators of this warning sign:

- Lack of a systematic audit process
- Lack of reporting
- Management unaware of these systems' status
- Status not reviewed on a set frequency
- Viewed as part of normal operation
- No mitigation plans or concern when systems are out of service (OOS)
- No priority in maintenance planning system
- No check that systems are returned to service after being repaired

Some items to consider regarding this warning sign are given below.

- How was the issue brought to light?
- Is finding an OOS emergency standby system to be treated like an incident?

8.2.3 Poor process hazard analysis action item follow-up

Process hazard analyses are effective only if the action items identified are followed-up and closed. Without this step, the risk remains in place as if the PHA were never performed. This also can lead to or further compound the perception that PHA action items are not important or if low priority. Some aspects of this sign are as follows:

- The tracking system for PHA action items is not transparent, easily viewed, current, audited, or worthy of commitment
- Poor PHA action item follow-up practices, specifically in terms of detail, completion of all related subtasks, and meeting the set deadlines
- PHA items not easily tracked
- No central database
- Action items not assigned with due dates

Questions to ask include the following:

- What is required to help action item owners understand the importance of implementing PHA action items in a timely fashion?
- Is our tracking system at fault?
- Do PHA teams and action item owners need special coaching?

- Does leadership understand the importance of completing this part of the PHA process?

8.2.4 Management of change system used only for major changes

Management of change incorporates all true changes, not just those involving the installation or replacement of major equipment. Does the management of change system include process changes, organizational changes, material substitutions, removal of equipment from service, and other required items? Each of these minor changes has resulted in major incidents when neglected.

- MOC is viewed as an engineering department administrative process rather than an analysis of the proposed change for hazards.
- The MOC item number is used as the project number.
- The MOC process is not used for minor changes.
- Incidents have occurred in the process where changes were made outside the MOC process.

Here are some questions to consider:

- How are we helping MOC system users to identify changes more easily?
- What are we doing to help our teams avoid using management of change as an administrative process?

8.2.5 Backlog of incomplete management of change items

There are often two types of incomplete MOC packages. One includes post-startup approved activities such as painting which need to be completed later. The second includes pre-startup activities such as procedures or as-built drawing markups which may not be complete. This warning sign shows a critical flaw whenever an incomplete MOC package of the second type exists for a change that has undergone startup and is in operation. It is the ultimate sign that the management of change system has failed.

- How well does your site monitor the management of change system and its synchronization between pre-startup and post-startup activities?
- Do you audit past MOC packages for MOC item closeout issues?

8.2.6 Excessive delay in closing management of change action items to completion

Management of change item closeout incorporates the change in permanent management systems. Ineffective, lengthy closeout of MOC items results in increased risk. Changes could have been made in the field without the proper documentation updates and as-built drawings. Operators are in a position of operating without current accurate information. This can result in errors being made. Some indicators of this warning sign follow:

- Significant delay in closing out management of change items

- Allotted time from implementation to closeout is excessive
- MOCs implemented without ever being closed out
- Incidents occur where changes were made and operators do not have current information

Try asking these questions with your MOC system users:

- What are the primary root causes of the delays?
- How can site leadership better provide resources?
- Which specific resource categories are lacking most?

8.2.7 Organizational changes not subjected to management of change

This warning sign has been receiving industry's attention lately. A well-running process safety management system can be thrown into disarray by not analyzing the impact of organizational changes. These can be events such as:

- New employees or newly assigned employees
- Internal shifts of employee job assignments
- Exiting employees
- Reorganization of the facility leadership roles
- Changes in staffing levels
- Laboratory support
- Technical support

Some sites have moved key people without reassigning the critical process safety duties they fulfilled. Here are some questions to ask concerning this warning sign:

- Do you have a special section in the management of change administrative procedure to address the increased risks of any organizational change?
- Are administrative positions left open for long periods?

8.2.8 Frequent changes or disruptions in operating plan

The operating plan, a summary of the important goals for the process, gives overall guidance. Frequent changes can result in confusion and in lack of clear roles, responsibilities, and actions. Incidents may occur. People will not know what to do in abnormal situations or will lose reliance on the plan and set their own agenda. Indicators include the following:

- The operating plan is being affected by unreliable operations.
- Maintenance outages are canceled.
- Maintenance is reactive, not planned.
- Operation continues with equipment in a poor state of repair.
- Incidents can be traced to poorly managed operating plan changes.

- The operating plan is focused on a production number and not on aspects of total operational excellence. That is, the plan should be based upon performance excellence in safety, environmental, quality, and economic success.

Here are some questions to consider:

- Can the operating plan be more clearly defined?
- What drives the maintenance prioritization, and how does it relate to the operating plan?

8.2.9   Risk assessments conducted to support decisions already made

The practice of performing the risk assessment after key decisions were already set into motion is not a practice that is open to assessing new risks. This practice implies a paper exercise to capture what has already been determined. This warning sign's indicators are as follows:

- There are only a few decision makers in the organization. They tend to make decisions by themselves based on their limited experience.
- Decisions are clearly predetermined and implemented prior to risk assessments.
- Risk analysts are under pressure to provide the right number, not the correct analysis.

If this warning sign is present, consider these issues.

- Is this a matter of organizational culture?
- How do you communicate that this practice is detrimental to the organization?

8.2.10   A sense that *we always do it this way*

If persons involved in management of change and risk analysis find examples that deviant behaviors are becoming normalized as described in Chapter 1, *Introduction*, this warning sign is present. Some indications of this warning sign follow:

- Process parameter(s) have shifted to a new steady state but are still within acceptable norms.
- Signs and tags have become unreadable, but everyone knows what they are.
- Comments are heard such as…*He is already a trained operator for process B; we will give him a quick overview of process A, and then he can operate it as well.*
- Comments are heard such as…*That relief valve always sticks; just thump it once a shift and it is ok.*

- Operators find out about changed equipment weeks after the change was made.
- Procedure changes needed due to physical plant changes are not made.

These behaviors all represent the acceptance of a broken process safety management system.

Here are some questions to consider:

- Do you audit for these mostly perception-based warning sign indicators?
- How does the site's leadership team address the issues, and how effectively?
- What is the perception of the facility staff concerning this warning sign?
- Has there been an initiative to validate current practices and ensure correction of out-of-date practices?

8.2.11 Management unwilling to consider change

Management support is critical to the success of process safety. If management appears unwilling to accept change of all types, they may stifle needed improvements, or there could be shortcuts that do not follow the process. These actions may result in undesired actions and incidents. Indications of a management resistant to change include the following:

- Workers perceive that their input is not needed or welcomed.
- Management resists change and wants to continue in the same path.

Here are some questions to consider:

- What direction is the management team getting from the level above them?
- Do the organization's executives understand the business case for process safety?

8.2.12 Management of change item review and approval lack structure and rigor

When the staff of a facility sees the MOC process as an activity-based formality, a warning sign is present. A management of change approval comes with accountability to the change. The management of change process is diluted if accountability is not present. The human pieces of the equation are essential for due diligence, and their positive attitude toward the change system is critical. Other indicators are as follows:

- The management of change process is too flexible. It is not prescriptive enough regarding who must approve starting up the change after all prerequisite actions are met.
- The system is able to bypass key decision makers because it slows down the process or they ask too many questions.

- There is no evidence of feedback, questions, or occasional rejection of MOC items.
- If there is a fast-tracked emergency MOC option in your site management of change element, you find that it is used quite often. This could be a sign that persons who are more interested in getting changes installed than in managing them appropriately misuse the emergency MOC option.

Consider discussing these issues:

- Is the current MOC system being abused or bypassed?
- Was the system designed to help ensure process safety excellence?

8.2.13 Failure to recognize operational deviations and initiate management of change

If robust procedures or processes are not in force, operators can migrate into a thought process where they may not easily recognize a deviation from normal operation or the attending consequences. Analyze these deviations so that if an event occurs, the worker evaluation methods, training processes, and procedure documentation are in place to address the situation.

- Are operating decisions based solely on experience without the added benefit of current analysis?
- Is there a general lack of evidence that appropriate procedures, processes, and guidelines exist?

8.2.14 Original facility design used for current modifications

The facility design should be challenged through a process hazard analysis (such as a HAZOP study or other method) to confirm that all risks and changes from the original design are identified using accurate as-built drawings and process information. If process safety efforts have been neglected and modifications not subjected to MOC, there could be many unmanaged changes lurking in the process. Further modifications need to be managed by a rigorous MOC process.

Here are some questions to consider:

- Are the as-built drawings accurate?
- Has a plant capacity test run been conducted to gauge throughput capability?
- Are there other benefits from engaging in such an activity?

8.2.15 Temporary changes made permanent without management of change

This warning sign is the same as if a change had slipped through the system. A section of the site's management of change program document should set stringent methods for knowing the status of all temporary changes and reevaluating each of those changes later considered for permanent status.

- Does the MOC procedure require a second review prior to making a temporary change permanent?
- If you already have such a system, is it working?
- Do temporary changes continue to be renewed to avoid formal review and approval?

8.2.16 Operating creep exists

Operating creep, like organizational creep, is a subset of the general term *normalization of deviance*. Operating creep is the state where process conditions change over time without being managed while remaining within safe upper and lower limits. These operations may be moving closer to an unsafe condition without recognition of the danger. Periodic review of current operating parameters against established operating limits would help address this risk.

- Have you checked to determine if operating creep is an issue for the processes at your site?
- Has operating creep caused your normal operating parameters to exceed the original design ratings?

8.2.17 Process hazard analysis revalidations are not performed or are inadequate

Once a formal hazard evaluation is complete for a process, it needs periodic update and revalidation to reflect changes. Not only is this a good practice, it is a requirement of regulation in many countries. As we discussed earlier, identify hazards before evaluating and managing the associated risks.

Early in the life of a process, the critical thinking abilities of the PHA team may be limited to the chemicals present and the process itself. It is over time during the operation of the process that this thinking is improved. Revalidations allow review teams to apply lessons learned from incidents and near misses that have occurred since the original, or previous, PHA and provide the team with a holistic view of all MOCs that have been generated since the last review.

Here are some questions to consider:

- Is there an industry benchmark for PHA revalidation management?
- Is there a benchmark for PHA revalidation quality?
- How many re-do PHAs did the site perform after the initial total process PHA was complete?

8.2.18 Instruments bypassed without adequate management of change

Is there a difference of opinion among the technical staff as to the need to control situations where environmental health and safety critical devices (EHSCDs) are removed from service? If critical instruments require maintenance or calibration, risk increases for the period when they are out of service. Appearance of this warning sign may be related to organizational culture. Revisit those warning signs to see if a connection exists.

Many sites have an administrative procedure or section of their management of change system that helps ensure effective analysis, authorization, communication, and recordkeeping for disabling environmental health and safety critical devices for brief out-of-service periods. That method can always be used for managing EHSCD shutdowns if desired. However, the steps described in such a procedure can provide an effective and streamlined approach for managing these specific types of changes. A full management of change review is generally required if the outage goes beyond a few days. One approach to consider is described below.

- Determine the device or system's EHS criticality.
  - o Note that disabling an EHSCD when the shutdown is specified by an existing approved operating procedure (for example, bypassing an SIS to enable safe startup or shutdown), the approved operating procedure for that activity applies.
  - o Note that this approach does apply to disabling an EHSCD when the shutdown is specified by an existing approved mechanical integrity procedure or work instruction.
- Determine the effect and risk impact of a planned activity.
- Describe disabling an EHSCD for preventive maintenance.
- Describe disabling an EHSCD for upsets, unexpected shutdowns, or equipment malfunction.
- Complete the disabling EHSCD approval form and associated documentation.

Often the procedure includes a table of acceptable mitigation measures as minimum guidance for the staff to add appropriate layers of protection during instrument outages.

- Have you evaluated the site's methods for controlling instrumentation bypass situations?
- Are there improvements that can be made?

8.2.19 Little or no corporate guidance on acceptable risk ranking methods

Once hazards have been identified and the risks associated with them have been analyzed, the acceptability of the risk should be determined. Management should communicate its risk tolerance criteria and its expectations as to how to apply them. This is essential to help guide hazard identification and risk analysis teams in deciding whether the hazards identified warrant further actions or controls to achieve the company's risk goals. It also helps ensure that they select the most appropriate risk control measures. An incorrect perception of risk could lead to either inefficient use of a company's limited resources or the unknowing acceptance of risks that exceed the risk tolerance of the company.

The risk tolerance criteria and risk analysis procedure should be documented with the appropriate level of detail to ensure that risks are prioritized and communicated in a consistent and accurate manner. Written objectives should clearly state the benefits to the company and express the benefits in terms that demonstrate the values of these processes.

- Has there been a failure to communicate a company's risk tolerance and document related procedures and programs? This is the main indicator of this warning sign.
- Does management follow the company's risk tolerance criteria when making decisions affecting safety or production?

8.2.20 Risk registry is poorly prepared, nonexistent, or unavailable

Once risks have been identified, it is a common practice in a risk management program to enter them into a risk registry. A registry of all significant risks in the workplace is an important reference tool, as it can assist in identifying new hazards in other areas and provides a record of actions taken to control existing risks. Failure to update or, in the worst extreme, failure to prepare a risk registry may indicate a company's inability to follow through on the results of the hazard identification and risk analysis efforts. This makes risk-based decisions difficult.

A risk registry allows these results to be communicated to management so that they understand the risks that have been identified and the actions proposed to manage them. A registry also allows a company to effectively manage recommendations and track actions to completion through the development of action plans that establish responsibilities and implementation deadlines.

Finally, a risk registry allows a company to reevaluate its residual risks periodically. The registry helps a company keep these risks on the front line. This makes it a straightforward task to compare the existing risks to any new risks or to evaluate the possibility of further reducing risks as new technologies and programs become available.

- Is your company's risk registry updated on a regular basis?
- Is the risk registry presented to management as part of the regularly scheduled process safety stewardship?

8.2.21 No baseline risk profile for a facility

A baseline risk profile represents a facility's established risk based upon completed risk studies and past risk decisions. It is a benchmark for making future risk decisions. A risk profile portrays all types of risk including safety, environmental, production, reputation, and assets. If a baseline risk profile has not been developed, it may suggest that management does not really understand risk. It also might suggest that process hazard analyses, facility siting, and global issues may not be adequately managed.

- Is your facility's baseline risk profile up to date?

- Is the facility risk profile in compliance with the corporate risk acceptance criteria (or risk matrix)?
- Is an effort made to ensure that future changes do not deviate from the established baseline profile?

8.2.22 Security protocols not enforced consistently

Security measures are required at industrial facilities to ensure that unauthorized persons do not gain access to hazardous processes and equipment. Unskilled and unauthorized persons put themselves and others at risk when they enter a hazardous facility. An individual with criminal intent (such as a disgruntled employee) can cause an event of significant consequence if allowed to enter a facility based on familiarity or facial recognition. Vehicles should also be searched to ensure that unauthorized materials do not enter a site.

There may be a casual tendency on some sites to admit workers based on their work attire or the logo on the vehicle they are driving. This should be prohibited. In addition, safety and security shortcuts should not be taken for visitor tours, audit teams, or upper-level management. Also, workers should not into the facility allow fellow workers who may have forgotten their badges.

- Is your facility surrounded by a fence or barrier preventing access to unauthorized persons?
- Are warning signs posted advising visitors of the hazards and of the requirements for access?
- Are all workers required to display a company-issued ID badge complete with photograph?
- Are access privileges canceled when a worker terminates employment?
- Are all vehicles entering your facility routinely searched upon entry?

8.3 CASE STUDY – CYCLOHEXANE EXPLOSION IN THE UK

June 1, 1974, the Nypro (UK) site at Flixborough was severely damaged by a large explosion (see figure 8-1). Twenty-eight workers were killed and an additional 36 injured.

The explosion resulted from a cascade of events that began in March. On this date a vertical crack over 6 feet long in cyclohexane reactor No. 5 was discovered. The facility was subsequently shut down for an investigation.

It was decided to remove the No. 5 reactor and install a temporary bypass pipe. The bypass was not designed by a qualified piping engineer. The operation was restarted on April 1. The bypass line lifted off the temporary supports but initially held its integrity.

Two months later, on June 1 during another startup, the bypass line, connected at either end with bellows, deformed in a V-shape and failed. The resulting release of cyclohexane formed a vapor cloud that found an ignition source and exploded.

What went wrong, and how could they have seen the signs?

Figure 8-1   The Nypro Explosion Aftermath

The investigation into this incident concluded the following critical facts:

- There was an insufficient management of change process.
- Normalization of deviance occurred (cyclohexane leaks).
- Hazard identification of changes within the facility was poor.
- The assembly was not pressure tested.
- Full calculations were lacking.
- Maintenance procedures for recommissioning (reinstatement) were inadequate.
- Facility layout and control room design did not recognize the possibility of major disaster occurring.
- Operating procedures were not robust.
- A hazard review should have been performed.

There was a significant time delay from the discovery of the problem to the major incident. Meetings were held, decision makers were involved, and action plans were created and implemented. They recognized that there was a serious problem. Company officials and people involved probably would have said they were performing due diligence. Dutch State Mines (DSM) determined the direct cause of the reactor leak, and it is common to bypass equipment to maintain production. This is something most companies should face, and generally, it is done successfully. Prior success does not equate to good systems or good control.

There was inadequate management of change.

The initial vertical crack in the reactor was due to using nitric acid to adjust the pH of the water to control leaks. This addition of nitric acid resulted in nitrate stress corrosion cracking. Nitrate stress corrosion cracking was a well-known phenomenon to metallurgists. This was a failure in managing change.

The investigation determined the root cause but did not stop this practice. It also did not extend into management systems or how the change was made, approved, put into service, how the water control was monitored, or whether testing was required.

The leaks were early warning signs that something was not in control. The crack was an indication that management of change was seriously flawed. Outside experts identified the root cause, but follow-up action did not resolve the issue.

Safety, engineering, and technical reviews should occur prior to facility and process modifications. These reviews should be traceable and identify proposed changes to process conditions, operating methods, engineering methods, safety, environmental conditions, engineering hardware, and design.

Would a management of change process alone have prevented this occurrence? It would have helped, but management of change processes must be robust enough to manage rare, catastrophic events. If there is no management of change process, this has a serious potential. Moreover, if there is a management of change process, it needs to be reviewed for effectiveness to determine if it will meet its functional requirements. Adequate time should be allowed for a complete hazard identification analysis and documentation. Does the site identify potential catastrophic events and train people how to react? People involved should know their roles and responsibilities and have appropriate skill sets. An analogy is pilot training for emergencies. Would your facility have the same rigor, calmness, and reaction to events if it was an airplane at 30,000 feet and the left engine flamed out?

There was inadequate recognition of the process hazards.

An early warning sign was the frequent cyclohexane leaks. Leaks were routine events. They occasionally resulted in a process shutdown but did not appear to trigger a full-scale analysis of the full root causes and how to prevent them. Leaks represent a loss of containment. They are serious events that are not acceptable and require recognition and preventive action. The installation of the bypass line did not give due regard to the dangers of cyclohexane.

All employees must be aware of the hazards associated with the work they carry out and be able to determine that the risks involved are acceptable. In this case, the hazards did not receive adequate review. There was inadequate technical expertise at the site and the review was focused on how to restart and not on management system failures.

Audits are a key tool used to recognize management system failures. Third-party audits that take a cold-eyed review of systems can be particularly useful. Audits examine how things are *meant to operate* in comparison to how they are *actually* operating. Thorough audits allow action items to be uncovered without relying on learning from incidents.

The hazard analysis should be carried out to determine the possibility and consequences of the hazards. A form of risk assessment should then identify what hazards have been created by the change that may affect process or personnel safety and what action can be implemented to reduce or eliminate the risk. Additional hazards that may be introduced which need to be considered are fire, explosion, and loss of containment.

Hazard identification should be carried out by the right people in the proper environment with the right information and the right process with the correct focus and support from senior management.

Management systems and how decisions are made should have processes created around them so the right results occur. These should be in place prior to an incident and robust enough to withstand production strains and other pressures.

The line was installed without a pressure test, stress analysis, or adequate inspection.

The failure of the bellows, twisting of the line in service, and inadequate supports were highly suspect in the subsequent investigation on the prime failure mode.

The warning signs were that the line was installed without the proper hydrostatic pressure test. It was discussed and canceled due to estimated difficulty. There appeared to be no consideration for blinding at the reactor vessels and installing high point vent and drain. An in-service test was chosen, but it is unclear what checks were done to confirm good operation. The line lifted off the temporary scaffolding supports and twisted with the applied load.

A proper review of the bellows, supports, and stresses in normal and abnormal operating conditions would have revealed that the bypass line was not suitably designed for the service.

A hydrostatic test with appropriate inspection points and design considerations would have indicated potential issues with the design. It would have shown that the line was potentially unstable and moved unacceptably with the applied load.

There were no stress calculations done or alternative equivalent protection methods proven. The calculations performed were to prove that the line had the flow capacity.

The early warning sign is that a very dangerous action was taken without a plan to monitor it in service and determine what actions would be taken based on certain observations.

- Piping design standards were not checked.
- The facility did not have the expertise and did not go off-site to obtain it.
- There were no drawings made. The hazards of the process and consequences of failure were not factored into the analysis.
- The line was installed on temporary supports. These supports were not properly engineered.
- The line was supported on temporary scaffolding supports, and when put in service, lifted off these supports.
- When the line lifted off the supports, it did not trigger any action or further analysis. At this point, the reactor nozzles were supporting the line, and essentially the bellows were carrying and reacting to the full load.
- The bellows vendor manual was not checked for the designed service details.

The facility layout and inventory did not recognize consequence of failure.

Eighteen people died in the control room. There were no survivors. The facility did not have an adequate hazard analysis performed on its initial design.

- The facility layout did not recognize that a major event could occur.
- The control room was not designed to withstand a major event.
- The facility inventory was not minimized, and there was inadequate ability to isolate the process in case of a major event.
- The principles of inherently safer design were not followed.

Operating procedures were not written to the level of detail suitable for the hazards.

The explosion occurred during startup when abnormal situations occurred that required modifications to the startup phase operating procedures. Leaks were occurring, nitrogen was not available, the reactor had to be placed into dry cycling, the temperature did not stabilize during the dayshift, and the reactor could not be vented to lower pressure. Eyewitness reports indicate that some incident had occurred 30 to 60 minutes prior to the main explosion. These are all signs that things were going wrong.

Sites should have written procedures, with adequate and thorough reviews, for abnormal events. This includes procedures for shutdown. Operators should be trained on these procedures.

The procedures should draw reference to previous incidents, their causes, and means of preventing them. Preventive measures (process control, instrumentation, protective measures such as containment, reactor venting, quenching, and reaction inhibiting) should be addressed.

The procedures for critical operations should use best practices with sign-off on steps and responsibilities clearly identified. Hazards should be included in the procedures with action steps.

Maintenance processes failed.

Two loose bolts were found on a critical joint. Small failures can result in major failures. Reinstatement checks and sign-offs are required. Incidents on less critical lines would have given warning that their systems were not robust enough. Operator walk-downs should have occurred before systems were put back in service to ensure that bolted joints are tightened and the right gaskets installed. Procedures and the entire maintenance process should ensure that maintenance staff is aware of the hazardous nature of the working conditions.

Well-written maintenance procedures consider human factors, the skill level of the workforce, maintainability principles, failure rates, fault recognition criteria, and marginal performance.

Exercise: Can you identify other warning signs that may have preceded this incident?

9
AUDITS

To measure is to know. If you cannot measure it, you cannot improve it.
Lord Kelvin

9.1 AUDITS SUPPORT OPERATIONAL EXCELLENCE

Standards and management systems define process safety program requirements. Audits measure compliance with those requirements. Although audits are often resource intensive, they are the best method for measuring a facility's improvement over time and for comparing facilities with similar risks and regulatory profiles to determine how to use resources effectively to improve programs and performance. An effective audit program is one of the best defenses against complacency. However, an ineffective audit program can waste valuable resources and yield misleading information, which may suggest that the facility is better off than it really is. Effective audit programs keep organizations focused on the significant issues and result in long-term improvement and sustainability. Effective audit programs keep senior management informed about compliance with organizational and regulatory requirements, and allow them to make risk-based decisions on how best to allocate both capital and personnel resources. Effective audit programs also measure senior management's commitment to process safety.

9.1.1 Audit team characteristics

Audits are team activities. Therefore, team composition is critical. Each audit team should include members with the following characteristics:

- Subject matter experts in the audit focus areas
- Knowledge of the regulatory and organizational requirements that apply to the facility being audited
- Basic understanding of process safety and how the activities within process safety are interrelated
- Basic understanding of the protocol being used
- Basic training in auditing and interviewing techniques

- At least one member of the audit team should know the basic operations of the processes being audited. Consider including process operators.

The audit team leader should:

- Have attended a process safety management training course, either in-house or a generic public course offered by a reputable training organization
- Be trained in audit team leadership
- Have participated in previous audits as a team member

9.1.2   Internal and external audits

Audits can be internal or external. Internal audits can occur within the facility itself with no outside assistance. This is often termed a first-party or self-audit. Another form of internal audit is an internal second-party audit. Internal second-party audits involve the assistance of personnel from other facilities or the corporate office. Internal second-party audits may be led internally, or by external resources. Internal second-party audits are often effective because they encompass outside points of view and experience, and can be very helpful in sharing best practices between sites. External third-party audits involve an outside party in the audit process. This external third party can be a contractor, regulatory agency, or customer. External third-party audits often involve assistance from facility personnel but are usually led by the external third party.

9.1.3   Audit protocols

The audit protocol defines the scope of the audit and gives structure and guidance to the audit team. Some audit protocols are general in nature and measure the overall functions of the facility management system. Some protocols are very specific and focused, and measure some aspect of compliance at a granular level. Many protocols today are designed for auditing integrated management systems, and may measure compliance from a safety, environmental, and quality perspective concurrently. Although the integrated audit approach is often the most efficient way of measuring a facility's compliance from multiple perspectives, it is often less focused on specific regulatory requirements and can give the facility and audit team a false sense of security. Take care when you conduct integrated management system audits to ensure that sufficient detail is given to satisfy specific regulatory requirements.

When conducting audits, the audit team is focused primarily on these major questions:

- Does the facility have a policy or procedure for the required activity?
- Does the policy or procedure adequately satisfy the specific requirements of the protocol?
- Are accountabilities defined?

- Does the practice in the field demonstrate compliance with the policy or procedure?

The audit team should use a number of methods to measure compliance. This should include a detailed inspection of policies and procedures for adequacy, verification through inspection of documentation records, and validation through interviews of facility personnel that the activity is occurring at the desired level to satisfy compliance.

9.1.4   Addressing audit results

Audits are measurements of compliance. The auditing process will determine (1) if the facility has a policy or program to address the specific requirement, (2) if the policy or program is adequate to satisfy compliance, and (3) if the facility has a corresponding practice that is compliant with the policy or program. While conducting the audit, the audit team will observe both compliance from these three perspectives and areas where improvement is needed. The organization should determine if they will recognize both noteworthy activity and areas for needed improvement. Some organizations focus only on deficiencies, and do not recognize noteworthy activities that were observed during the audit. However, many organizations have determined that recognizing truly noteworthy behaviors is critical for ensuring that the activity will continue.

When audit findings are determined, the audit team will formulate recommendations needed to improve compliance. These recommendations will be translated into action items and an action plan. Monitoring and management of these action plans are usually the responsibility of the facility. However, in many organizations the corporate function has become more involved in monitoring the progress of these action plans to completion and reporting progress on a regular basis to the organization's senior management. Everyone involved in the audit function, from the audit teams to facility management to the organization's senior management should be aware of how monitoring action item status and the resulting progress reporting will be accomplished. Surprises in these functions are seldom productive.

9.2   THE AUDIT RELATED WARNING SIGNS

These are the warning signs associated with the auditing functions:

- Repeat findings occur in subsequent audits.
- Audits often lack field verification.
- Findings from previous audits are still open.
- Audits are not reviewed with management.
- Inspections or audits result in significant findings.
- Regulatory fines and citations have been received.
- Negative external complaints are common.
- Audits seem focused on good news.
- Audit reports are not communicated to all people affected.

- Corporate process safety management guidance does not match a site's culture and resources.

9.2.1   Repeat findings occur in subsequent audits

When subsequent audits result in repeat findings, one of the following occurred:

- The facility did not act upon the original finding.
- The facility did not understand or agree with the original finding.
- The facility developed an action plan related to the original finding, but the action plan was not adequate to resolve the issue, or the action plan was not adequately implemented.
- After the facility implemented the action plan, the noncompliance reappeared.

When recurrence occurs, the organization might ask the following questions:

- Is our action plan monitoring system adequate?
- Is there proper management accountability to ensure that action items are resolved effectively and in a timely manner?
- Are there systemic process safety problem areas that need more frequent or different types of monitoring?

Repeat or unresolved audit findings are often signs of an inadequate audit process or an inappropriate follow-up and monitoring system.

9.2.2   Audits often lack field verification

The first step in managing risk is to identify the need for a policy or program, and develop the documentation. However, implementation of the program or policy requires communication and training. Often this is the most difficult step in the actual implementation of the plan.

In an attempt to improve audit efficiency and reduce the intrusive nature of the process, the audit team often spends too little time verifying that systems are in place and working at the point where the activity is being performed or the control is needed. This results in a paper-based audit. Similarly, audit findings need to be supported with field verification if a credible case is to be presented to facility management. Negative biases alone do not constitute an effective audit.

To determine if actual field verifications are inadequate, ask the following questions.

- Have we tried to reduce the actual audit time by reducing the amount of time we spend in the field verifying that systems are in place and working?
- Does the audit team have enough team members with the appropriate level of expertise to ensure that field verification efforts are effective?

- Is there an appropriate feedback process that allows field verifications to affect the audit process?
- Is there at least one team member with enough facility experience to know where field verifications should be occurring and by what method?

Once programs have been developed, field verification of systems that are in place and working is one of the most important steps of the audit process.

9.2.3 Findings from previous audits are still open

Often, audits will reveal compliance issues that require capital expenditures and significant time to correct as well as non-conformances that are more easily addressed. When audits reveal findings from previous audits that have not been resolved, a number of questions need to be asked to determine the significance of this warning sign.

- Why is the finding still open?
- Has an action plan been developed to address the finding?
- Is the facility tracking the progress of the action plan on a regular basis?
- Have the action plans been prioritized?
- Is site (and corporate, if appropriate) senior management knowledgeable and supportive of the action plan?
- What are the possible consequences of not closing the open findings in a timely manner?

There is a significant difference between an incomplete action plan that has been prioritized and had resources allocated to complete the action items based on an assessment of risk, and an audit finding where no action plan has been developed.

9.2.4 Audits are not reviewed with management

Accountability for process safety performance rests with the senior manager of a business unit or facility; such accountability requires effective performance feedback. When it is determined that findings from previous audits have not been adequately addressed, either through audit system monitoring and follow-up or through subsequent audits, ask the following questions:

- Was the original audit closeout meeting conducted with appropriate members of management in attendance?
- If a senior manager could not attend an audit closeout meeting, did a qualified delegate attend in their place? If so, did that person adequately communicate the proceedings to his or her superior?
- Were copies of the audit report sent to management, and was an opportunity provided to answer questions pertaining to the findings?
- Does the facility have the resources to address the finding adequately?

- Does the organization (or facility) have a formal review process downstream of audits to ensure that findings are addressed in a timely manner?
- Are the results of audit reviews and monitoring adequately communicated to senior management on a regular basis?
- When deficiencies are noted during audit reviews and monitoring, does senior management respond in a manner that indicates a high level of commitment to the importance of the auditing process?

When audit results and subsequent follow-up and monitoring of performance are not reviewed with management, there may be a deficiency within the system itself, or a failure of management's commitment to the importance of the auditing function.

9.2.5   Inspections or audits result in significant findings

Audits are time-consuming activities that require resources in both time and personnel to complete. However, every organization would rather find a significant nonconformance through an audit than as a result of an incident investigation after an accident. The more often a facility or function is audited effectively, the less likely it is that significant findings will be found. Occasionally, however, audits will uncover situations that are significant. When this warning sign occurs, ask the following questions:

- Is the finding something that has always existed and is being discovered for the first time?
- What allowed this significant finding to develop?
- To what extent have management of change shortcomings contributed to this significant finding?
- Is this finding a systemic issue (one that may now exist in other areas of the facility and/or organization)?
- How are significant findings communicated throughout the organization so that everyone can learn from the event?
- Has a change in the audit program or audit personnel contributed to the discovery of the finding?
- Has a causal analysis been conducted to determine the root cause of the finding?

9.2.6   Regulatory fines and citations have been received

How are fines and citations viewed within your organization? Are they recognized as a failure that requires corrective action, or are they considered as part of the cost of doing business? The number, amount, and severity category of regulatory fines or citations can be a warning sign. It is relatively easy to compare similar sites' performance related to governmental inspections with your organization's performance. It is often useful to know where your organization's performance is relative to your industry's average. Ask yourself the following questions about your organization's approach to regulatory compliance:

- Is there a process to ensure that all required actions associated with the fines or citations are implemented in a thorough and timely manner?
- Is regulatory compliance regularly reported to senior management?
- Does your organization incorporate regulatory compliance checks in your internal audit program and implement corrective actions before citations occur?
- How does your organization effectively learn from similar experiences (both internally and externally) and react to regulatory issues prior to incurring a citation or fine?

If your organization has received an unusually large fine or a large number of smaller fines, inadequate management commitment to regulatory compliance or lack of an effective compliance management system may be indicated.

9.2.7 Negative external complaints are common

Negative media reports related to audit performance often indicate two possible issues. First, a regulatory audit resulted in a high number of citations, serious violations, or large monetary fines. Second, the organization's social responsibility is questioned by the community. Ask the following questions:

- Is your organization taking positive corrective action to resolve the findings?
- Are external complaints taken seriously, and are they reported within the organization in a transparent manner?
- Do negative reports from external sources cause your company to re-evaluate its operations, or are they considered fringe comments and likely to be ignored initially?
- Does your organization regularly invite the community to enter into a dialogue with the facility?
- Does the site support and communicate with local emergency planning committees and first responders?
- Is an effort made to communicate progress on such issues back to the source?

9.2.8 Audits seem focused on good news

Are audits conducted where there are few, if any, negative findings? Does your organization constructively review its audit performance looking for ways to improve? If an audit report seems to promote the strengths of the process safety management program implementation over the weaknesses inappropriately, it may be indicative of an atmosphere where the audit team does not feel comfortable reporting areas for improvement. This may have happened explicitly, by facility management asking that negative findings be tempered or removed from a draft audit report. It may have happened when outside third-party auditors were trying to maintain a beneficial working relationship with an organization for economic reasons. No matter the influence or reason, auditors should not alter their impressions unless missing evidence or new evidence is provided during the draft review that supports changing their initial findings.

- Are the audit team's findings edited by management to downplay potential problems?
- Are audit teams debriefed after an audit to determine if there were any inappropriate influences?

Audits are detailed inspections of systems and processes, looking for opportunities to improve performance. When best practices are observed, they should be noted and reported. However, when audits are unusually positive they may be indicative of an organization that is not willing to receive constructive criticism and is not willing to admit there are areas where it needs to improve.

9.2.9   Audit reports are not communicated to all affected employees

Are audit findings considered confidential and only available to management or a select few? All facility employees deserve to know the status of the organization's performance. Audit results can help motivate employees to participate in management system activities. If the organization's process safety performance is good, employees deserve to be recognized for their participation and help in the compliance efforts. For organizations with multiple sites, sharing audit results provides an opportunity for sharing of best practices. The following questions can help you determine if this warning sign might apply to your organization.

- Has your organization established a standard process for sharing audit results to all affected employees?
- Is the status of action plans associated with audit follow-up regularly communicated to all affected employees?
- Does your organization support the audit function by inviting employees to participate in the audit process?

9.2.10   Corporate process safety management guidance does not match a site's culture and resources

Organizations with multiple facilities and business lines have risks specific to each site and the processes involved. Facilities vary greatly from a cultural and resource perspective. In an effort to achieve a sustainable process safety program within large organizations, standards and guidelines may be either too comprehensive or ambitious for all sites to comply. Sometimes the opposite is true, and organizations establish guidelines and directives for the lower-risk or lower-resourced facilities, and do not adequately address the needs of the larger, higher-risk sites. If your organization is large and global in nature, consider the following questions:

- Have you developed high-level guidelines that are broad in nature and allow sites flexibility in defining specific site programs tailored to the specific site needs?
- Have you included representation from the sites to be involved in the development of organization-wide programs?
- Do you have a system for sharing best practices among all facilities within the organization?

Although most process safety programs ultimately have the same goal, the most effective programs are those that are tailored to the specific risks and needs of individual sites.

9.3   CASE STUDY – CHEMICAL WAREHOUSE FIRE IN THE UK

A large chemical warehouse was built adjacent to an existing Allied Colloids chemical facility in Brad, England, in 1989. Although the primary purpose was to protect freeze-sensitive oxidizing agents from cold weather, over 400 other industrial chemicals were stocked in the warehouse. Documentation provided with the facility included a chemical segregation plan and safe work practices for handling of chemicals. Of the 125 workers assigned to work in the warehouse, not even one was qualified to work with chemicals. A major fire occurred at the facility in 1992 and much of the warehouse was destroyed. Thirty workers and emergency responders were hospitalized. Considerable environmental impacts were incurred because of contaminated firewater runoff. If the fire had continued to spread, a large quantity of acrylonitrile and methyl chloride might have been released or consumed. The main production area of the facility could also have been destroyed.

On the morning of the incident, steam heated blowers in the warehouse had been turned on to dry out moisture. At least one of the stored chemicals was heat sensitive. The heating caused some drums of heat-sensitive material to rupture and spill white powder on the floor. A passing worker believed that the powder was smoke and raised the alarm. No action was taken, but efforts were made to retrieve a chemical data sheet. Within a short period, the material ignited and a fire quickly spread through the warehouse and into the facility. There were no firewater sprinklers installed in the warehouse. A total of 2700 tons of chemicals were consumed in the fire. After a full hour had passed, emergency response teams arrived, but no firewater was available to fight the fire. A contingency plan involved piping water from a distant reservoir. This further delayed the response. A large quantity of contaminated water entered the sewer system, and fish were killed in a nearby river. It was later determined that the powdered chemical probably encountered unattended spills of persulfate and other oxidizing agents. The mixture is likely to have been ignited by an impact, possibly from a lid and associated metal ring closure from one of the damaged drums falling onto the floor. The fire spread throughout the warehouse and smoke was blown toward nearby roadways. The fire was tended for 18 days to reduce the risk of a subsequent ignition and flare-up.

This incident was the result of a total breakdown in management systems. It started with the assignment of an unqualified administrator to build and manage the warehouse. Problems continued well into the operation of the facility. Voids and deficiencies were apparent in every element of process safety. Previous audits, including at least one by the Health and Safety Executive (HSE), had identified many of the issues that contributed to the incident, but no action had been taken to address these concerns. Lack of follow-up on audits and inspections is a significant warning sign that can have tragic consequences.

Exercise: Can you identify warning signs that may have preceded this incident?

10
LEARNING FROM EXPERIENCE

The only real mistake is the one from which we learn nothing.
John Powell

10.1 METHODS FOR CONTINUOUS IMPROVEMENT

Within the risk based process safety framework, the learning from experience pillar evaluates both failures and successes in all aspects of a business. Every success and every setback is an opportunity for organizational learning. Whether we learn from a proactive measure or from a response to an error, we need to determine what worked well and how we can ensure that this experience is integrated into our work practices and procedures. In the case of setbacks, we need to know what went wrong and how we can avoid repeating the same errors. This may be the most difficult challenge. The biggest error that an organization can make is in missing these opportunities. What do we need to do differently to maintain a healthy sense of vulnerability even when we are performing at our best?

10.1.1 Incident investigation

The site process safety programs detailing *incident reporting, investigation, and follow-up* should combine to create an incident learning and prevention work process. Ensure that near miss reporting and investigation are emphasized and provide feedback to workers at regular intervals to encourage this practice.

- Build documentation and tracking into an existing action item tracking system.
- Analyzing historical incident data for physical causes, immediate causes, and management system failures (root causes) can be very useful in preventing future incidents.
- Formal incident investigations should attempt to identify warning signs that might have existed prior to the event.

149

10.1.2 Measurement and metrics

A facility administrative document on *process safety metrics* typically provides tools for evaluating a facility's process safety performance and gives examples of performance measurements. It can describe audit reporting methods and how to track continuous improvement in process areas that have significant risk levels. Additional process safety metrics can be stewarded through an employee safety committee. Process safety incidents as well as leading and lagging indicators should be tracked and communicated to corporate and hourly personnel.

10.1.3 External incidents

Around the world, incidents occur in all industries. Whenever humans interact with complex machinery and significant hazards, there is potential for something to go wrong. Some of these occurrences result in catastrophic events with associated media attention and public concern. Industrial trade associations attempt to communicate incident reports to their member companies to help them learn from other members. Governmental and private concerns provide details on incidents that have occurred elsewhere through databases and published reports.

There is a tendency among some plant managers and employees to dismiss the experience of others on the basis that *it cannot happen here.* In fact, many facilities that have experienced catastrophic losses had good occupational safety records and considered themselves to be relatively safe from disaster. The same mistakes made by one organization, regardless of industry sector or type of facility, may be occurring within your own organization. Can you afford to take that chance? Even large corporations have not experienced all possible incidents, nor can they afford to do so. Take advantage of the opportunity to learn from the mistakes of others.

10.1.4 Management review and continuous improvement

The site's process safety program should define individual responsibilities for specific activities tied to each of the elements of process safety. Management reviews should be formalized and scheduled at regular, planned intervals. To augment this and maintain an awareness of current activities, management should tour the facility and visit workers in the field.

The concept of continuous improvement suggests moving forward. To do this requires a reliable reference point. Otherwise, mistakes of the past will be repeated and little gain will be achieved. A set of company archives should be maintained and reviewed periodically to ensure that the organization continues to move forward.

To paraphrase Trevor Kletz, a renowned expert on process safety, "Organizations have no memory." Unless positive steps are taken to avoid loss, even serious incidents will recur. In simple terms, the things that caused one incident are just as likely to cause another.

10.2 THE LEARNING FROM EXPERIENCE WARNING SIGNS

These are the warning signs associated with the failure or inability to learn from experience.

- Failure to learn from previous incidents
- Frequent leaks or spills
- Frequent process upsets or off-specification product
- High contractor incident rates
- Abnormal instrument readings not recorded or investigated
- Equipment failures widespread and frequent
- Incident trend reports reflect only injuries or significant incidents
- Minor incidents are not reported
- Failure to report near misses and substandard conditions
- Superficial incident investigations result in improper findings
- Incident reports downplay impact
- Environmental performance does not meet regulations or company targets
- Incident trends and patterns apparent but not well tracked or analyzed
- Frequent activation of safety systems

Many of these warning signs are chronic in nature. They are easy to ignore and may be accepted as the norm. Insufficient effort or ineffective fixes may have been attempted repeatedly, but the problems were not resolved.

10.2.1 Failure to learn from previous incidents

Failure to learn is often evident by the fact that operations tend to repeat the same mistakes. Can you think of industrial sites that have experienced major catastrophes, only to be revisited by similar or related incidents in a relatively short period? This warning sign may suggest the need for a complete process safety management upgrade.

Every process safety incident that occurs at a facility has at least one associated cause. The precise lineup of causes determines the outcome of the incident. Unless the causes are systematically analyzed and addressed with follow-up actions, there is a finite probability that the incident will recur. In fact, under slightly different circumstances the incident may have consequences that are more serious. Process safety gaps and deficiencies are sometimes difficult to detect and even more difficult to correct. A major incident provides an opportunity to explore the root causes and take action. It also represents an opportunity to transform a site or company culture in a fundamental way. If this opportunity is overlooked, nothing will really change and the operation will remain vulnerable to repeat losses. If this warning sign is present at your facility, consider the following questions:

- Does your organization reevaluate the entire process safety system periodically to find opportunities for improvement?
- Are significant learning opportunities from investigations reviewed with the workforce, and are these integrated into company training?
- Is there a written archive of significant incidents and associated learning?
- Is there a training initiative to enhance understanding of process safety philosophy and practices at the facility?

10.2.2 Frequent leaks or spills

A high frequency of leaks, spills, or resulting minor fires (or even *smolder* events) indicates that the primary goal of handling hazardous materials in a safe and controlled manner has not been met. Containment has failed. Small leaks, spills, or fires often have minimal consequences. When these events occur repeatedly, workers can begin to lose their sense of vulnerability. Moreover, this situation may represent normalization of deviance where leaks and fires are considered normal as long as no one was hurt. In this environment, workers and management can start tolerating increasingly severe leaks, spills, and fires. In some cases, releases of materials such as corrosive liquids and vapors can degrade the integrity of surrounding equipment, resulting in unexpected equipment failures. Ultimately, a major release or fire can result. Some specific indicators that leaks, spills, or minor fires are occurring with high frequency may include the following:

- No tracking or trending of minor incidents (that is, there is no way to even know if such events are occurring)
- Tracking exists and shows that such events are occurring but no action is taken
- Small leaks or spills known to exist but not fixed or stopped in a timely manner
- Widespread evidence of spills and leaks, such as staining or corrosion at leak points
- Incipient stage fires (that is, fires that can be extinguished easily by one person with a portable extinguisher or hose) occur but are viewed as a necessary part of operations
- Small leaks meant to be temporarily patched are left in service for long durations

If these events are occurring with high frequency, ask these questions:

- Is a robust incident tracking, trending, and investigation program in place that captures relatively minor incidents?
- Are trending reports reviewed and acted upon?
- Are incident investigations related to leaks or spills collectively analyzed to find commonalities?
- Have you evaluated the recommendations and their implementation plans?

- Have resources been assigned to identify weaknesses to reduce loss of containment incidents?

10.2.3   Frequent process upsets or off-specification product

Maintaining a process within its safe operating range for the maximum period is a sign of mastery in operations, maintenance, and engineering. If operators are frequently responding to process upsets (including *nuisance alarms*), if process data reveals that processes are frequently operating outside normal limits, or if off-specification product is common, then the process is not operating or being maintained in a controlled and consistent manner. This situation is a form of *normalization of deviance*. It may also represent inadequate process design. These warning signs can result in the following situations:

- Operators become so conditioned to responding to process upsets that they stop viewing them as potential safety events and do not respond in a timely manner. Eventually an upset may be significant enough to exceed safe limits, but operators may not recognize it and may not respond appropriately.
- Off-specification product may be a result of upstream design or operation problems that may have potential safety ramifications if not corrected.
- If operators are subjected to frequent nuisance alarms, they may begin to treat alarms that are more critical in the same manner.

If frequent process upsets or off-specification events are present:

- Are process upsets (or alarms) tracked and analyzed? Are corrective actions increasing in frequency?
- Are upsets treated as incidents (or near misses) and investigated?
- Has the number of potential nuisance alarms been minimized?
- Are product quality events tracked and investigated to determine the root cause?

10.2.4   High contractor incident rates

A high rate of contractor-related injuries and incidents could indicate trouble in one or more of the following six areas:

- Contractor selection
- Availability of appropriately skilled contractors
- Contractor training
- High contractor turnover
- Contractor safety auditing
- Contractor performance evaluation
- Contractor oversight

Contractors often are intimately involved in operating and maintaining hazardous chemical processes. Presence of this warning sign may indicate that your contractors do not have an understanding of hazards, well-defined procedures, appreciation for safety concerns, or sufficient oversight and control of their workers. When dealing with hazardous chemicals or processes, any of these concerns readily can escalate to conditions ripe for a potential catastrophic incident.

If this warning sign is present at your facility, consider the following questions:

- Does the facility have a contractor management program?
- Do you evaluate contractor incident reports for completeness, and then identify the areas above that could benefit from redesign or reemphasis?
- Do you review your contractor training and awareness tools and ask questions, such as:
  o What do the contractors need to know?
  o How do we inform them of hazards?
  o What drives contractors to perform their duties safely?
  o How does our contractor safety system compare to those of other companies where our contractors work?

10.2.5 Abnormal instrument readings not recorded or investigated

If abnormal readings are not being recorded or addressed, a serious problem is occurring which can lead to complacency, inability to analyze an impending critical situation, human error, and potential disaster.

Address each instance on a risk-based priority and determine the cause of failure. Action items should focus on prevention across the site. If this warning sign is present at your facility, consider the following questions:

- Do you question out-of-range or abnormal readings at your facility?
- Are the instruments reporting properly, or are they defective? Some abnormal readings could indicate deteriorating process conditions. Defective gauges, sight glasses, and instruments are not acceptable.
- Are operators more prone to acknowledge or suppress the alarms and not respond or bypass the instrument or submit work requests to investigate or repair the instrument?
- Have clear criteria been established for identifying instrument abnormality issues that require investigation?
- Have you applied the necessary resources to address repetitive upset conditions caused by abnormal readings?
- Is there controlled access to the control system and alarm set points?

10.2.6 Equipment failures widespread and frequent

Trends in repetitive equipment failures indicate a flawed management system for addressing equipment integrity. Such failures can range from seemingly minor to relatively significant. Some examples may include the following items:

- Continual gasket leaks or failures
- Continual valve packing or pump seal leaks or failures
- Pumps or motors that frequently need repair
- Equipment (for example, an air compressor) that is constantly tagged *out of service*
- Pipe or equipment supports breaking or showing significant corrosion

This warning sign creates a definite increase in the probability of experiencing a catastrophic incident. If present at your facility, consider the following:

- Are equipment failures formally investigated?
- Is equipment integrity limited to the selection of equipment, or does it also address the way that equipment is operated and maintained?
- Is the necessary expertise present or available to develop and implement a robust equipment integrity program?

10.2.7 Incident trend reports reflect only injuries or significant incidents

If non-injury incidents or near misses are not included in regular site incident trend reports, it suggests that the organization is looking only at a narrow subset of incidents, many of which may not be process related and not be looking at the bottom of the catastrophic incident pyramid for leading indicators. By looking only at injuries and other lagging indicators, an organization is avoiding looking at the data that could have prevented the incident or that could prevent catastrophic process events. If this sign is present, consider the following questions:

- Have we defined all the useful data that the management team needs for safety-related decision making?
- Do we track and trend process upsets or activation of safety systems even if no serious consequences occurred?

10.2.8 Minor incidents are not reported

Whether due to fear of reprisal, bad attitude, lack of incentives, or time restraints, if employees sense management acceptance to not reporting incidents, there is a problem. A low level of operational discipline exists throughout the site. Workers sense that management does not care about safety in spite of all those posters to the contrary. Not reporting an incident should be an actionable job performance issue. The organization should support these actions at all levels. When this warning sign is present, the organization has minimal or no ability to identify learning opportunities and take critical actions to prevent the same incidents from recurring. Consider the following questions if this warning sign is detected:

- Do you have a system for promoting incident reporting?
- Has the workforce been trained on the importance of reporting certain categories of minor incidents?

- How quickly and effectively does your site respond to incident or safety deficiency reports? Nothing reduces thorough reporting faster than employees who see no management interest or response.

10.2.9   Failure to report near misses and substandard conditions

This warning sign can reveal an organizational culture problem. When employees do not report obvious physical hazards, spills, or near miss situations, safety management loses an entire segment of vital data. If this warning sign is present at your facility, consider the following questions:

- Have you determined the need for a more rigorous practice of encouraging near miss and incident reporting?
- Has the workforce been trained on near miss reporting?

10.2.10 Superficial incident investigations result in improper findings

In today's resource-limited world, it can be easy for a facility to fall into the bad habit of performing only superficial incident investigations, as well as generating weak or inappropriate findings and corrective actions. This can also result from a push to investigate more incidents within an organization that has a poor process safety culture. Regardless of the source, the organization does not achieve the goal of determining root causes and generating effective corrective actions. Evidence of this warning sign may include the following items.

- Root causes were not determined, and only the incident symptoms were addressed.
- Root causes are often labeled as *equipment failure* but do not address the underlying reasons for the equipment failure (for example, installed improperly, specifications inadequate, maintenance and inspection insufficient, or other details).
- Root causes avoid identifying *management system* issues as causes and instead focus on worker error or equipment failure, which are both contributing causes.
- Action items from an incident investigation are so narrowly focused that they only address a very specific piece of equipment or a specific step in a process.
- Action items are applied only to the specific equipment or unit affected. The applicability at other units or other similar equipment at the site is not considered.
- Action items are too general, such as *improve training, follow procedures, improve maintenance.*
- Action items are similar from incident to incident, for example, repeated use of *improve training* or *provide more training.*
- Management focuses on and monitors the number or count of incidents investigated but does not monitor the quality of the investigations.

If this warning sign is present, it can mean that we are not taking the critical actions necessary to prevent recurrence of an incident. Consequently, the incident may occur again. Of greater concern is a similar preventable incident with significantly more severe consequences. When this warning sign is present, consider the following questions:

- Do you regularly evaluate the effectiveness of the site incident investigation procedure?
- Does the site procedure include means to assess incident investigations periodically for quality and thoroughness? Reporting deficiencies to management might help drive corrective actions to improve the process.
- Does the procedure clearly specify how incidents are to be investigated, and does it require identification of root causes?
- Do you train workers on incident investigation techniques before (or as part of) involving them on an incident investigation team?
- Do you audit previously performed investigations to learn where the deficiencies were? Rushing investigations to resume production or to avoid the pain of another reminder that our systems are not functioning may suggest a cultural issue.

10.2.11 Incident reports downplay impact

It is essential that the incident-reporting database include a category for identifying and documenting the consequences of process safety incidents. In addition, incident investigation team leaders who use the system need to ensure that we record any incident that may have been a precursor to a catastrophic incident.

Note that the presence of this warning sign or the three immediately preceding it is particularly worrisome. It indicates that the organization is essentially unable to see to its process safety problems.

- Do you provide training for workers who participate on an incident investigation team to focus special attention on any incident categorized as a process safety incident?
- How do you communicate the causes and potential outcomes clearly?
- Do your facility's first-line supervisors or team leaders feel comfortable writing a report that reflects negatively on their work team?
- Do your facility's first-line supervisors or team leaders hesitate to write a report that they believe will not be acted on?
- Has the environmental, health, and safety department or leadership team escalated communication to a level where an effective corrective action can be initiated?
- Have you evaluated incident trends over the last three to five years for commonalties? This exercise can be eye opening.
- Are your investigation teams unbiased?

10.2.12 Environmental performance does not meet regulations or company targets

Good environmental performance is a direct reflection of operational excellence. Although we typically manage environmental-related activities outside the process safety arena, environmental incidents are often the result of loss of containment. The CCPS business case determined that the actions taken to manage process safety have direct business and environmental benefits. Conversely, if an operation or facility cannot meet its environmental commitments (either company targets or local regulations) it is difficult to comprehend how things might be better in the process area. Process safety requires a 100% commitment toward preventing loss of containment from closed systems. Any indication of weakness suggests the potential for a process safety incident.

This warning sign indicates a possible commitment to standards without the associated resources to implement needed engineering and operational changes. If a site's *license to operate* depends upon meeting specific release limits, repeated violation is especially critical. Could site culture be an issue? Is this a simple matter of installing necessary equipment and controls?

- Have environmental performance issues that might require incident investigation been clearly defined?
- Has your organization applied the necessary resources to address repetitive exceedance conditions?

10.2.13 Incident trends and patterns apparent but not well tracked or analyzed

When an organization expends the energy to collect incident data and then does not analyze it for signs of future problems, this warning sign is present. It represents a waste of human resources and capital by not determining the potential leading indicators that could be revealed by such analysis. It also shows a lack of awareness of what is really going on within a facility. More important, failure to analyze trends suggests that the organization does not fully understand loss prevention principles and does not think they are important at the supervisory or management level. Few incidents happen outright. Visible indicators and trends that are readily identifiable through analysis precede most incidents. A trend from the norm is the first step toward establishing a new baseline, which in turn signifies normalization of deviance. Without trend analysis, an organization cannot set priorities or make sound business decisions. If this warning sign is present at your facility, consider the following questions:

- Has a set of metrics been established for measurement and analysis on a set period?
- How do you include incident trend analysis in regular management stewardship?

- Do you identify physical, procedural, and organizational changes that might have coincided with incident trends and determine if these were contributors to incidents?

10.2.14 Frequent activation of safety systems

Safety systems such as safety instrumented systems (SISs), mechanical shutdown systems, and pressure relief devices are critical safety measures. These devices are often the last line of defense in preventing the release of a toxic, reactive, flammable, or explosive chemical. Thus, they are important barriers in the prevention of catastrophic incidents. Activation of these systems, including excessive alarming, may indicate that a process is operating significantly outside its normal limits and potentially at or beyond safe limits. In some facilities, however, activation of these systems may be viewed as part of normal operations (that is, they feel *it did what it was supposed to do*) rather than an incident. This is a process safety culture issue. Frequent activation of these systems may indicate that a normalization of deviance is occurring. This situation should trigger an immediate investigation to determine the causes of the activations.

Some sites use safety system challenges as a leading indicator for process safety performance. If workers and management do not view activation of such devices as safety incidents, it is unlikely that anyone will notice an upward trend in the frequency of activation. In addition, no one will be investigating these events and identifying corrective actions. The causes of these events will remain unchecked. This represents normalization of deviance and can result in accepting these potentially unsafe actions and conditions as normal.

When this warning sign is apparent, it suggests that the facility may be operating at excessive levels and beyond nameplate capacity for some of the equipment. Consider the following questions:

- Has your facility defined the type of safety system challenges that require incident investigation?
- Do we emphasize the importance of tracking activations of safety systems for loss of primary containment protection and require that such events be investigated?
- Do we communicate to workers that activation of such safety systems represents a deviation from normal and potentially even safe operating limits?

10.3 CASE STUDY – SPACE SHUTTLE *COLUMBIA* INCIDENT IN THE UNITED STATES

In February 2003, the U.S. space shuttle <u>Columbia</u> disintegrated during reentry into the Earth's atmosphere, killing all seven crew members.

The loss of <u>Columbia</u> was a result of damage sustained 81 seconds after launch when a piece of foam insulation the size of a small briefcase broke off the space shuttle external tank (the main propellant tank) under the extreme launch forces. The debris struck the front edge of the left wing, damaging the thermal protection system. The mission continued but ended tragically 16 days later when the orbiter attempted reentry.

NASA's original shuttle design specifications stated that the external tank was not to shed foam or other debris. Potential strikes upon the shuttle itself were safety issues that needed to be resolved before a launch was cleared. Launches were often given the go-ahead as engineers came to see the foam shedding and debris strikes as inevitable and irresolvable, with the rationale that they were either not a threat to safety or were an acceptable risk. The majority of shuttle launches recorded such foam strikes and thermal tile scarring. During reentry of mission STS-107, the damaged area allowed the hot gases to penetrate and destroy the internal wing structure rapidly, causing the in-flight breakup of the vehicle.

Mission STS-107 was the 113th space shuttle launch. It was delayed 18 times over the two years from its original planned launch date of January 11, 2001, to its actual launch date of January 16, 2003. A launch delay due to cracks in the shuttle's propellant distribution system occurred one month before a July 19, 2002, launch date. The Columbia accident investigation board determined that this delay had nothing to do with the catastrophic failure six months later.

The Columbia accident investigation board's recommendations addressed both technical and organizational issues. It was established that the failure of the foam insulation had resulted in an impact with a critical wing component. However, the organizational contributors to the accident were far more complex. Among these were normalization of deviance; denial of vulnerability; ignoring opinions of safety staff; lack of consistent, structured approaches for identifying hazards and assessing risks; and worker intimidation, contributing to communication breakdown.

The Columbia accident was unique in terms of the chain of events that led to the wing incident. However, the management system failures strike a closer resemblance to other major accidents and in particular, the Challenger accident that occurred 17 years earlier and the Apollo 1 fire in January 1967 that killed three astronauts. One of the revealing statements from the investigation report states: "In our view, the NASA organizational culture had as much to do with this accident as the foam." Professor Andrew Hopkins, a renowned process safety expert, had this to say about the incident: "[The decisions] were not made by the engineers best equipped to make those decisions but by senior NASA officials who were protected by NASA's bureaucratic structure from the debates about the wisdom of the proposed actions."

Several quality and reliability problems were apparent to staff in the two years prior to the incident. Shedding of foam and debris became common events that were deemed inevitable. Although potential impact incidents on the shuttle were initially viewed as safety concerns that could delay a launch, the tolerance for these increased with each successful launch. The main safety focus appeared to shift to the propellant distribution system, which had experienced cracking.

The space shuttle program was still operated as a *business* and had to compete with other programs for federal funding. NASA officials focused on the success of each mission and discounted hazards and risks that were more apparent to technical staff. Technical experts had serious concerns about recurring impact incidents but were unable to communicate these to the decision makers. In such a situation, people (management) tend to hear what they want to hear. Once again, normalization of deviance breeds the belief that *we are beyond reproach and invulnerable.*

The Columbia accident was one of great disappointment. Had the recommendations from the Challenger accident in 1986 been fully implemented, the systemic issues that resulted in the loss of Columbia could have been avoided. Accident prevention need not be a complicated activity. Look no further than your last incident to make substantial gains and improvements.

Had management engaged in an effective dialogue with workers on the project, it is likely that the true risk might have been perceived and acted upon. Can you suggest other things that should have happened as a result of prior experience at NASA?

Exercise: Some of the warning signs that may have been noticed prior to this incident are listed below.

- Conflict between production goals and safety goals.
- Incident reports that downplay impact.
- Failure to learn from previous incidents.

Can you identify other warning signs that may have been present?

11
PHYSICAL WARNING SIGNS

Another flaw in the human character is that everybody wants to build and
nobody wants to do maintenance.
Kurt Vonnegut, Jr.

11.1 THE EVERYDAY THINGS MATTER

Physical warning signs are often tangible and visible to the naked eye. However, we sometimes see only what we expect to see. Workers should be on continuous lookout for these early signs of failure. It will be difficult to notice them if they have become a normal part of doing business. This section is closely linked to the CCPS concept book *A Practical Approach to Hazard Identification.* Its location is near the end of the book to emphasize in earlier chapters the importance of looking for systemic failures. For many workers, the material in this chapter should present less of a challenge than the material presented in earlier chapters.

The presence of physical warning signs suggests that the standard of acceptance has dropped to a low level, which indicates that deviations have been normalized. Some of the physical warning signs are as simple as housekeeping. Others may indicate major problems. The first step in mitigating physical warning signs is obvious. They must be noticed. Moreover, our *intuition* may indicate that something is wrong.

If it seems unsafe, it probably is!

These physical warning signs should be prioritized for rapid follow-up whenever found.

11.2 THE PHYSICAL WARNING SIGNS

These warning signs are often obvious. Practice looking for them the next time you walk through a facility. Customize the list for your processes.

- Worker or community complaints of unusual odors
- Equipment or structures show physical damage
- Equipment vibration outside acceptable ranges
- Obvious leaks and spills
- Dust buildup on flat surfaces and in buildings
- Inconsistent or incorrect use of personal protective equipment
- Missing or defective safety equipment
- Uncontrolled traffic movement within the facility
- Open and uncontrolled sources of ignition
- Project trailers located close to process facilities
- Plugged sewers and drainage systems
- Poor housekeeping accepted by workers and management
- Permanent and temporary working platforms not protected or monitored
- Open electrical panels and conduits
- Condensation apparent on inner walls and ceilings of process buildings
- Loose bolts and unsecured equipment components

11.2.1 Worker or community complaints of unusual odors

A facility's license to operate may depend upon strict control of air, water, and waste emission sources. Whenever unusual odors (or well-known odors that should not be detected) are present, it represents a possible containment breach or failing equipment. Loss of containment can result in fire, explosion, or toxic release.

- Do you have a system to identify the source of the odor carefully using the appropriate monitoring equipment and techniques?
- If appropriate, do you perform an incident investigation?
- Does the organization effectively implement action items to correct each problem (whether a leak, electrical failure, or friction damage), evaluate the response, and modify the administrative or engineering layers of protection to prevent future similar occurrences?

11.2.2 Equipment or structures show physical damage

Physical damage indicates a weakness. For whatever reason, equipment and structures that are damaged may not be fit for service and could be vulnerable to further failure. The fact that such equipment has not been removed or repaired is a concern. When permanent equipment (for example, pumps, vessels, piping, and pipe supports) shows obvious structural damage or wear, it requires a quick response. The response may range from installing a temporary barricade or barrier, to analysis and reporting, to increased inspection frequency, or to a unit shut down for immediate repair. Nevertheless, perform an analysis of the risks and a resulting response. Equipment damage can lead to loss of containment and result in fire, explosion, or toxic release. Structural damage of handrails or other safety structures due to rust or corrosion can lead to a fatality.

- Has the problem been identified?
- Have you performed an incident investigation if appropriate?
- If so, did they implement action items to correct the deficiency, evaluate the response, and modify the administrative or engineering layers of protection to prevent future similar occurrences?
- Has the organization simply forgotten that structural components are not designed for misalignment and that response is urgent in most cases?
- Are personnel trained (or retrained) to improve their hazard recognition skills? This training could use simple walkthroughs with an experienced person, or training materials with pictures showing equipment in its proper state.
- Are there related or similar processes within the facility that need to be inspected quickly for similar damage?

11.2.3 Equipment vibration outside acceptable ranges

When rotating equipment or structural equipment is experiencing excessive vibration, the equipment is under stress. Equipment design specifies operation within an acceptable range of vibration. For high-speed rotating equipment with close tolerances, even seemingly minor imbalances can cause severe damage. Vibration can also be caused by flow and phase change phenomena. Vibration can cause damage to equipment or piping and cause welds to fail. This can lead to loss of containment and result in fire, explosion, or toxic release.

- Is your organization skilled at identifying abnormal vibration and the specific causes of the vibrations?
- Do you perform an incident investigation if appropriate?
- Will the site implement action items to correct the deficiency, evaluate the response, and modify the administrative or engineering layers of protection to prevent future similar occurrences?
- Does the facility have a predictive maintenance program that uses permanent vibration monitoring on critical pieces of equipment?

11.2.4 Obvious leaks and spills

Simply walking through a process area can reveal the existence of this warning sign. If there are noticeable spill remnants or stains around facility equipment or staging areas, or residual materials used to contain the spill are still there, it may reflect an organizational normalization of deviance. Missing or blistered paint or unexpected corrosion can indicate a leak of caustic or solvent material.

Prevalent leaks and spills are indicative of management system failures in one or more of the following five areas:

- Safety (both personnel and process), due to potential exposure to the material
- Environmental, by allowing the system to leak in the first place if the material is regulated
- Quality performance
- Maintenance, due to the failure to properly maintain equipment and materials
- Economics, due to the impact of the cost of the loss of product or rework, and the associated cleanup costs

This may be a cultural issue due to normalized deviance. Handle it at the source using the questions below.

- Does your organization identify the specific causes of the spills and leaks?
- Do you perform an incident investigation if appropriate?
- Does the site implement action items to correct the deficiency, evaluate the response, and modify the administrative or engineering layers of protection to prevent similar future occurrences?

11.2.5 Dust buildup on flat surfaces and in buildings

Dust buildup in an industrial facility is a common byproduct of cutting, grinding, and material handling operations. The accumulation of dust can present a significant fire or explosion hazard. For some processes, engineering solutions may be necessary. Excessive dust buildup may create a situation where the facility is at risk of a dust explosion. In fact, excess dust buildup conditions that were either not recognized or effectively managed preceded all major dust explosions. It is important to recognize that dust itself may not be the direct hazard. An ignition source and initial disturbance are often required to agitate the dust and create an airborne opaque phase capable of supporting the explosion. Since these initiating events are rare occurrences, many facilities have not encountered the secondary explosions despite having had a severe buildup of dust on building surfaces.

Any powdered materials building up on equipment, tables, workspaces, or building supports indicates a housekeeping issue. For some processes, we may need to engineer the solution due to the nature of the materials handled. The U.S. Chemical Safety Board (CSB) released a thorough study of industrial dust explosions early in 2006. Their study that showed from 1985 through 2005 there were 281 reported dust fires and explosions throughout industry in the United States. These accidents resulted in the deaths of 119 people and injured an additional 718. Research of news reports conducted by the *Combustible Dust*

Institute revealed that in 2008 in the United States there were over 200 dust fires and explosions. Basic good housekeeping practices reduce the risk of dust explosions greatly.

- Has the facility failed to establish housekeeping standards related to dust control?
- Do you monitor adherence to policies and redirect behavior using the organization's human resources system?
- Does the company identify the specific causes of the dust buildup?
- Do you perform an incident investigation if appropriate?
- Has the site implemented action items to correct the deficiency, evaluate the response, and modify the administrative or engineering layers of protection to prevent similar future occurrences?

Even one observation of dust buildup is a very significant warning sign. The presence of this warning sign often indicates an acceptance of deviance. It also suggests that the operation may lack a sense of vulnerability. If you see excessive dust buildup, ask these three questions:

- Is the facility aware of the dust hazards, and has the dust been tested for explosivity in a recognized laboratory?
- Is management aware of the potential risk of an explosion and, if so, what is being done to manage that risk?
- Why is there no sense of pride or ownership in the appearance of the area?

Even where a dust explosion is not imminent, dust buildup may have serious implications to an operation. It may be difficult to read instruments or inspect equipment covered in a significant dust buildup. Dust can affect a worker's health through irritation of mucous membranes of the eyes, nose, and throat. Dust may also enter equipment enclosures, contributing to friction, abrasion, and premature equipment failure.

11.2.6 Inconsistent or incorrect use of personal protective equipment

Every site has two categories of personal protective equipment (PPE). There is the standard category of PPE that workers are required to use at all times, such as protective footwear, eyewear, hearing protection, hard hat, and possibly flame-retardant clothing. Then there is the specialized category of task-related PPE. These are harnesses, respirators, breathing apparatus, chemical suits, electrical flash protection, and special gloves—all of the things that workers might use in a special situation beyond the standard PPE category. When you observe someone not using either category of PPE properly, it is a warning sign that employees and contractors are not adequately protected. It may also be a warning sign of poor training, inadequate supervision, or a poor safety culture.

- Do you periodically revisit the site safe work practices governing all types of PPE use?
- Have you performed a hazard assessment to determine the type of PPE required for the area or task?
- How do you monitor adherence to policies and redirect behavior using the organization's human resources system?

11.2.7 Missing or defective safety equipment

Safety equipment includes personal protective equipment, portable fire extinguishers, deluge systems, fire monitors, SCBA gear, or other special tools or devices needed to comply with safety policies. Since PPE is the last line of defense, damaged or misplaced PPE represents a significant hazard. Emergency response equipment is an important barrier in the prevention of catastrophic incidents.

- Does your site periodically revisit its practices and procedures governing all types of safety equipment storage, inventory, and inspection?
- Do you maintain a philosophy that emergency equipment is available 24/7 and if it is not, that a like in-kind replacement is in place?
- Do you leave sections of firewater piping blocked in for extended periods (> 1 day) because of leaks?

11.2.8 Uncontrolled traffic movement within the facility

Vehicle and personnel traffic must be controlled in an operating facility at all times. Commercial vehicles and large construction equipment can collide with sensitive process equipment, including overhead lines and pipe racks. The noise and vibration from such equipment can also interfere with instrumentation. Traffic congestion can impede emergency vehicle access. Often, the operators of commercial vehicles are not familiar with process hazards. Vehicle speed is another risk factor that should be controlled. Running motors are a potential ignition source and should be restricted in areas where flammable materials may be present.

Pedestrian foot traffic must also be controlled in process areas. Except for assigned personnel (plant operators), entry to a process area should be authorized.

- Does your site have traffic controls?
- Does your site have signs restricting vehicle entry posted at the perimeter of areas where hazardous chemicals are used?
- Are speed limits visibly posted?
- Are certain areas off-limits to vehicles because of size or weight restrictions?
- Are overhead pipe racks labeled with height clearance?

11.2.9 Open and uncontrolled sources of ignition

This warning sign can consist of noncompliant smoking, unrestricted vehicle entry, damaged process equipment, or failure to permit hot work activity properly. Many of these warning signs have preceded catastrophic incidents. The presence of an ignition source in areas where flammable chemicals are present can lead to an explosion.

- Do you periodically revisit the site safe work practices governing all types of process unit access, hot work, and smoking policies?
- Do vehicles drive into classified areas without a permit?
- How well do you monitor production areas to determine if new uncontrolled sources of ignition have developed?

11.2.10 Project trailers located close to process facilities

Recent catastrophic incidents have underscored the importance of facility siting issues. This is especially important for temporary structures. If project trailers are necessary for turnarounds or expansions, conduct facility siting studies to help ensure that they are located outside high hazard zones or primary chemical release paths.

- Does your facility have a policy or procedure that prohibits the erection or occupancy of project trailers in close proximity to hazardous process facilities?
- Is there a provision in your facility siting procedure that deals with temporary facilities?
- Are field inspections conducted to determine whether temporary facilities comply with the facility siting standard?

11.2.11 Plugged sewers and drainage systems

In addition to your desired finished products, processes often generate waste streams for your site to manage. Plugged sewers and drainage systems can prevent the safe removal of liquid wastes or byproducts from the process should an unexpected release occur. Flooding of the plot area may contribute to a fuel backup and subsequent pool fire. This warning sign suggests that basic maintenance and housekeeping have been overlooked. Do you inspect process areas frequently to determine if sewers and drainage systems are functioning properly?

- Are control methods in place to mitigate plugging after you identify it?
- Are measures in place to deal with local area flooding and sewer overflow?

11.2.12 Poor housekeeping accepted by workers and management

Whether seen in the control rooms at the consoles, in the break areas, or out in the facility, poor housekeeping practices can be one of the most important warning signs of system and cultural failures. Poor housekeeping is an issue seldom solved through one-time cleanup campaigns. Bad housekeeping practices are early warning signs of a culture that is normalizing deviance. Poor housekeeping also breeds lack of ownership. That becomes an organizational culture issue.

- Does the facility have regular area inspection programs where housekeeping performance is monitored?
- Are all workers in the facility involved in the area inspection and clean-up program?

11.2.13 Permanent and temporary working platforms not protected or monitored

You might see this warning sign on temporary scaffolding or on permanent facility structures. At a minimum, the facility should adhere to local safety regulations. Ignoring these basic and obvious means to prevent falls can indicate deeper neglect.

- How does the organization engage employees at all levels in a study to find and log substandard platform railings and gates?
- Does the organization identify substandard conditions and initiate immediate repairs?
- Does your site have a program to inspect permanent structures periodically?

11.2.14 Open electrical panels and conduits

Electrical panels designed for operation in a closed environment due to the presence of hazardous chemicals must be kept shut. Equipment inside electrical panels can be an ignition source and result in an explosion if hazardous chemicals accumulate. Leaving panels open, especially when located in hazardous service, indicates a poor process safety and housekeeping culture. If panel latches or sealing surfaces are defective, consider repair a high priority. Sometimes failures of this type are indicative of a *run-to-failure* mentality. This may be the result of an understaffed or under-resourced maintenance department.

- Are inspections conducted in critical areas to ensure that electrical panels and conduit are in proper condition?
- Do you repair noncompliant electrical panels and conduit quickly, and train workers on the importance of maintaining these important electrical components in good working order at all times?

11.2.15 Condensation apparent on inner walls and ceilings of process buildings

Detecting this condition in covered process areas or other ventilated workspaces can indicate inadequate or ineffective HVAC or workers leaving doors open for extended periods. Condensation issues may be trivial for some processes, but for others may create conditions ranging from product contamination, to corrosion problems, to air quality changes. In colder climates or with cryogenic operations, this can indicate that there are compromised insulation vapor barriers.

Condensation is less likely to occur when there is good air circulation within a building. Electrical area classification often depends upon a specified number of air changes per unit of time. When this condition is not met, the area may be in breach of its classification. Condensation provides a possible warning sign of this condition.

- If your buildings have experienced internal condensation, have such incidents been thoroughly investigated and corrected?
- Have you evaluated the engineering and administrative layers of protection for preventing condensation in work areas?
- Does the facility respond appropriately to correct substandard conditions in a timely manner?

11.2.16 Loose bolts and unsecured equipment components

As described for other physical warning signs, when a simple walkthrough reveals loose bolts, missing bull plugs, and improperly secured equipment, it can be an indicator that mechanical integrity is not a priority. The specific loose piece of equipment may not be an immediate problem, or it could involve low-risk equipment. However, the question implied is: *How many other pieces of equipment are in this state?* Loose bolts and equipment can pose a direct hazard to personnel and can contribute to large-scale incidents. Loose equipment can result in loss of containment and result in fire, explosion, or toxic release. When a component under pressure becomes detached, it may be propelled. This can cause harm to personnel or damage to sensitive equipment. Loose equipment is both a physical causal factor and a symptom of a weak process safety culture. Failure to recognize this condition has contributed to catastrophic incidents.

When assembling equipment, good practice requires certain checks to ensure that it is safe and reliable. This step should ensure that all components are secure. This includes piping and equipment support components. Over the course of a run cycle, equipment can loosen from stress or vibration. Operating organizations have a responsibility to check periodically that all equipment is tight and free from leakage or failure. The prevalence of loose equipment in an operating facility suggests lack of ownership. Furthermore, if loose equipment is visible to the naked eye, we wonder whether other, unseen equipment is secure. Such deficiencies are silent warnings of failure that could have serious consequences.

The presence of loose bolts and equipment in a facility suggests possible weaknesses in several management systems. Among these are process safety culture, operating procedures, asset integrity and reliability (maintenance), pre-startup safety review, and conduct of operations.

- Is the inspection of equipment for loose or mis-fitted equipment components part of the pre-startup safety review?

- Is operating equipment examined at regular intervals to ensure that it is tight and secure?
- When you identify substandard conditions, does repair occur in a timely manner?
- Is the proper operation and condition of latches, grating, and other components part of unit pre-startup safety reviews and associated checklists?

11.3 CASE STUDY – RESIN PLANT DUST EXPLOSION IN THE UNITED STATES

A dust explosion and fire destroyed a large section of the CTA Acoustics manufacturing facility in the southern United States in 2003. Seven workers were killed and another thirty-seven were injured. Part of the neighboring community was evacuated as a precautionary measure. This facility produced fiberglass acoustic insulation panels for the automotive industry as well as for other industrial clients.

The facility, built in 1972, initially manufactured insulation from fibrous cotton and phenolic resin. During the 1990s, the technology was modified and fiberglass was substituted for the cotton. This process change was completed in 2001. At the time of the incident, several similar production lines manufactured the acoustic panels. These production lines were located in close proximity to one another. Three raw materials—fiberglass, phenolic resin powder, and facing—were used to manufacture the panels. Main production equipment included material feeders, separating and mixing machines, curing ovens, and trimmers all connected by roller conveyors. The open area between some of the lines held storage racks of semi-cured acoustical insulation.

The following building construction details are relevant to the incident. The process equipment was enclosed in a structural steel frame building, consisting of vertical columns and horizontal beams. Exterior curtain walls, constructed of non-load-bearing prefabricated metal panels, were supported by and attached to the structural frame. The flat building roof was constructed of metal panels covered with weatherproof matting and supported by open-web steel joists, which were attached to the horizontal beams.

Different grades of phenolic resin powders were used on the various production lines, depending on the desired product specifications. Despite the existence of a bag house to remove residual dust from the process, a large quantity of dust had settled on the ledges of the process equipment and on the beams and panels of the building. Business continued as usual.

On the day of the incident, a temperature controller malfunctioned on one of the drying ovens. The oven door was then opened manually to cool the temperature inside the oven. Coincidentally, the bag house on that line had been operating inefficiently, creating excess dust in the facility. Efforts to repair the bag house and unplug a major blockage led to the creation of a dust cloud. The dust drifted toward the oven door and exploded. Although this explosion caused minimal damage, it shook all equipment within the building, disturbing dust that had settled on ledges and flat surfaces. A major dust explosion and fire soon followed, causing widespread fatalities and destruction.

There were several contributors to this incident. As was learned later, facility management personnel were aware of the dust hazards. However, they initiated no action, and the problem was not corrected. Visible dust hazards became commonplace and were simply overlooked.

- Are there obvious signs in your operation that chemical dust or residue is accumulating?
- Is there a cleanliness standard?
- Have responsibilities been assigned for ensuring cleanliness around process equipment?

Exercise: Can you identify warning signs that may have preceded this incident?

12
A CALL TO ACTION

Never mistake motion for action.
Ernest Hemingway

This book has highlighted numerous common and prevalent warning signs that have preceded noteworthy process incidents. Recognizing and responding to these signs is an important first step in reducing the risk of a catastrophic incident. A strong process safety culture should drive the commitment to process safety quality and minimize gaps and disparities. The previous chapters promote the belief that a strong process safety culture is a foundation to help ensure that we do important things when they need to be done. Such a culture respects the fact that things can go wrong and maintains a healthy sense of vulnerability within the organization. Avoiding catastrophic incidents begins with displaying a high level of operational discipline in your personal, team, and organizational behaviors.

Before developing an action plan based on the warning signs discussed in this book, consider the following quote by the British novelist Robert Brault:

Never act until you have clearly answered the question:
What happens if I do nothing?

The biggest obstacle to follow-up may be a sense of denial on the part of management. The longer a program defect or oversight has existed, the easier it may be to overlook its significance. Holding the belief that accidents only happen elsewhere commonly precedes many tragic incidents in industry. All workers have a role to play in influencing management that things need to improve if major incidents are to be avoided.

Although the desired effect of implementing an action plan is to improve our current situation, it does require us to change. The mere introduction of change in an organization can have ripple effects that should be considered. Changing the course of a large cruise ship takes time, monitoring, and a keen sense of direction. However, Captain Edward Smith and 1,517 people who perished on the *Titanic* would gladly have taken a different course if given a second chance on their way to the bottom of the chilled Atlantic.

Consider the Fédération Internationale de Football Association (FIFA) world football (soccer) championship. All teams that make it to the championship exhibit a high level of skill and experience. However, one team will survive all stages of elimination and go on to win the World Cup. That team maintained the drive and determination to win. In addition to high skills, that winning team exercised a high-level game strategy. This is analogous to the concept of *conduct of operations* in a processing facility. That term essentially means *do things right the first time* and *do not take shortcuts or accept compromises*. That is the distilled central theme of this book.

12.1 ACTIONS THAT YOU CAN TAKE NOW FOR EACH WARNING SIGN

While reading this book, you may have considered several ways to use the information to help your organization improve and recognize important warning signs. Consider some of the questions asked in this chapter. Step back far enough to see the interconnections among the various warning signs. Will one action in response to seeing one warning sign serve to forestall other warning signs? If your facility manager wears fire-retardant clothing to work most days, he or she probably has a basic understanding of the warning signs. Site leadership awareness is crucial in addressing them.

12.1.1 Periodic employee participation in analyzing warning signs

Begin using the warning signs as discussion topics for the regular safety meetings at your facility. When asking yourself if the warning signs listed in this book apply, or could apply to your organization, it is often helpful to use a team approach. Many organizations will involve managers, engineers, supervisors, and technicians in this process. This can be helpful in enhancing a healthy sense of vulnerability within your organization. Try the following technique:

- Introduce a warning sign or a category of warning signs.
- Invite open discussion from the participants on their opinion as to whether the sign is evident at your facility.
- Discuss and document the group's responses regarding how best to resolve the immediate issue.
- Discuss and document the group's responses regarding ways to prevent the issue from arising in the future.
- Identify effective approaches offered by the group for action.

These safety meetings should be lively and productive if the facilitator is skilled at drawing out participation and in smoothing differences of opinion. These discussions can also raise issues related to improving the organizational culture. Each warning sign discussed might result in a list of recommendations for system improvements.

Since 2002, CCPS has issued a monthly safety bulletin entitled *Process Safety Beacon*. Each *Beacon* highlights an important process safety theme and suggests things to look for in your operation. The *Beacon* targets operations and maintenance personnel in operating plants. This important application is one of several opportunities that complement the material in this book.

12.1.2 Use the warning signs as part of your next process safety audit

Consider adding a review of the warning signs as part of your process safety audits. You can modify Appendix A, *Incident Warning Sign Self-assessment Tool,* to assist in this. Many of the warning signs are dependent upon perception. However, larger sample sizes can help identify areas for improvement.

You can look at the associated warning signs for the element (or elements) under audit or you can examine them holistically throughout your organization. Review the resulting audit report with the facility management team. Focus on those warning signs found to be present. Remember to report on the activities your facility is succeeding at as well as the areas for improvement.

12.2 A SIMPLE PLAN TO CONSIDER FOR RIGOROUS IMPLEMENTATION AND FOLLOW-UP

If you find the concept of using warning signs as predictors of increased probability of potential catastrophe useful, you can build it into your process safety management system. Once your facility integrates warning sign detection and prevention methods into key elements of your management system, these methods will eventually become second nature.

12.2.1 Perform an initial warning signs survey

One of the most effective ways of using the information from this book is to perform an initial perception survey of affected employees. Identify the warning signs you would like to focus on using Appendix A, *Incident Warning Sign Self-assessment Tool,* as your survey tool. Use various methods to collect data as best determined by your site culture. Some methods are listed below.

- Surveys
- Interviews
- Document reviews
- Facility walkarounds
- Brainstorming sessions
- Accident database archive reviews
- Incident investigation analyses

Use a combination of methods that will work best to determine the status of perceived and measurable warning signs at the facility. This is your baseline. Any actionable determinations made during the survey should be designated as warning sign related action items for tracking and closure.

12.2.2   Build warning sign analysis into your management system

Warning signs are associated with the following areas of work within a safety management system. Should you modify your management system to include such items as the following?

- *Compliance audits*: Incorporate into your permanent audit protocols a check for the presence of warning signs.
- *Incident investigation*: Create a category of incident tied to warning signs. These can be tracked as a subset of near-miss incidents. Encourage the reporting of accidents and near misses.
- *Process hazard analysis*: Build warning sign analysis into all PHA revalidations as a validation point with the study team.
- *Process safety procedures*: When writing any procedure related to process safety implementation—operating procedures, maintenance task procedures, safe work practices, emergency response plan procedures— write the documents to avoid allowing warning signs to appear.
- *Employee training*: Consider adding a module on catastrophic incident warning signs and hazard identification to the new employee orientation-training curriculum and refresher-training curriculum for operations and maintenance employees.

You may find other opportunities to build the concept into additional elements. Consider revising the program and training the users on the changes as necessary. When thinking about your management system, consider the following questions:

- Do you have the measurements and key performance indicators (KPIs) that will be effective in helping you move your organization forward and measuring successes along the way?
- How do you ensure that desired practices are reflected in written procedures?
- How can historical data from measurements and KPIs be analyzed and trended to provide more insight and focus resources on the areas that need them most?

12.2.3   Use the new system and track related action items

Once your management system incorporates the facility's philosophy of using warning signs to enhance safety, track the effectiveness of this approach.

As day-to-day process safety related activities go on, make a special effort to monitor any action items from incident investigations, audits, or process hazard analyses related to addressing or avoiding a catastrophic incident warning sign. Evaluate each action item through to its closure. Consider monitoring the potential risks avoided, costs saved, cost of the action, and time to implement. Over time, the presence of these catastrophic incident warning signs should decrease within your organization.

12.2.4 Evaluate effectiveness in the next compliance audit

After incorporating the considerations of warning signs into your safety management system, the next periodic process safety audit can be a barometer of success. If the protocol includes the concept of using warning signs to enhance safety, you can categorize the audit findings to reflect their consideration.

Consider asking the following questions when determining effectiveness:

- How many of the warning sign related action items were closed on time?
- How many are still open or overdue?
- Were any warning signs reduced or eliminated in the period between initial survey and this audit?
- What were the costs of implementing the warning sign based action items?
- What is the state of recent key performance indicators? Have they changed since focusing on eliminating warning signs?
- Do you have a strong process safety culture?
- How do you measure whether it is well ingrained and displayed in facility systems, processes, and actions?
- How many new warning signs were discovered?

12.2.5 Maintain vigilance against recurring warning signs

Success in the area of process safety is a continuing challenge. Failure is ultimately measured by the occurrence of the types of catastrophic losses described in this book's case studies. Catastrophic incidents have defined some companies and been responsible for the demise of others. Failure to manage process safety almost always results in human pain and suffering.

Using the warning signs described in this book will be useful in helping improve your process safety performance. However, just because you feel confident that a warning sign is not present today does not guarantee that it will not resurface later. Most companies with strong safety cultures know that vigilance is important and that making periodic assessments to determine continued compliance is the only way to assure that process safety risks are being managed at an acceptable level.

Changes in management, changes in global markets, process changes, and changes in facility ownership are all factors that can affect whether a warning sign has an opportunity to reappear. Even companies we recognize as good performers from a business excellence perspective can have issues with these warning signs. Improved consideration of the warning signs does not mean they would go away forever. They can always reappear. Vigilance is required.

12.3 ACTIONS TO CONSIDER

Incorporating the facet of catastrophic incident warning sign analysis into your existing process safety or operational excellence management system can decrease the probability of major catastrophic incidents in your facilities.

Individual facilities should ask the following questions:

- How in-depth should we look at these warning signs?
- How good is our data related to the warning signs currently being exhibited at this facility?
- Can our process safety program benefit from adding the concept of catastrophic incident warning sign prevention?
- Is this essentially a type of predictive maintenance program for our business management system?
- How can we present the concepts in this book to higher levels of management?
- How can incorporating the considerations of these warning signs into our existing safety management system improve our program?
- What is the risk associated with not addressing these warning signs?

Initial warning sign survey results can be helpful in benchmarking various sites, or various process areas within a site. Organizations may want to consider using initial warning sign survey results from multiple process areas or sites to make risk-based decisions regarding process safety resource expenditures and upgrades.

The action required by the chemical processing industry at this juncture is to evaluate ways in which this concept of analyzing catastrophic incident warning signs can be used to enhance an operation's economic success, process safety success, environmental responsibility, and high quality standards.

Finding warning signs during perception surveys, audits, and other opportunities before loss events occur is more effective than looking for warning signs post-loss (that is, during incident investigations).

Ultimately, everyone has responsibilities related to process safety. Depending on your position in the organization, you may be involved in the development of the management system, the execution of the management system activities, or both. The following subsections include some of those responsibilities by organizational position, including senior management, managers, supervisors, and all employees.

12.3.1 Senior management

Senior management is responsible for knowing the risks posed by their operations, setting objectives to control those risks, and ensuring that action is taken. This includes potential risks to the environment, the community, and workers. Senior management also needs to understand how a catastrophic loss could potentially affect the business, both upstream and downstream of the operation. Avoid making excuses for any warning signs that are reported. There are no valid excuses. There are only reasons. Senior management should maintain a sense of vulnerability and communicate its support for risk-reduction and risk-management practices.

Senior management might ask the following questions:

- What is the present level of risk for each of the operations?
- Have I clearly communicated my support of risk-reduction activities?
- How can I promote risk-reduction efforts?
- Do we understand our personal safety ethic and motivation?
- Do we demonstrate critical safety-related leadership behaviors?
- How does our organization score regarding the culture factors that correlate to safety performance?
- Is our organization blinded by cognitive biases regarding process safety issues?
- Do we experience normalization of deviance?
- Is there an overemphasis on injury rates?
- How do our reward and recognition programs affect safety performance?
- Is our root cause analysis approach too simple?

12.3.2 Managers

Managers are responsible for taking the direction and leadership of senior management and changing the operation of the organization to meet the organizational objectives. Managers should promote the use of warning signs across the areas of their operation and make it a high priority through follow-up and demonstrated concern.

Managers might ask the following questions:

- Am I aware of the highest-risk operations in my plant?
- Is the emergency response plan adequate to mitigate the effects of a serious event?
- Is my staff aware of the warning signs in their areas?
- Are my operations effectively closing action items from audits, hazard assessments, management of change proposals, and incidents?

12.3.3 Supervisors

First-line supervision is one of the most important layers of leadership within an organization. Supervisors are the face of management to the employees and contractors. Supervisors can support or defy management's efforts and evoke an amazing response or ignore management's proposals with potentially devastating consequences. Most important, supervisors maintain the conscience of the plant during the time when other leaders are absent.

Supervisors might ask the following questions:

- Do I keep my eyes open for warning signs daily?
- Do I ask my employees to share concerns that could be warning signs?
- Is there a process safety message, lesson, or warning sign discussion in every crew meeting?
- Do my personal behaviors (speech, body language, attitudes, and actions) demonstrate my commitment to risk reduction?
- Do my employees hold me accountable for communicating warning signs to management?
- Do I communicate the *why* of the task along with the *what, how, and when?*

12.3.4 Using incident warning signs for operations leader training

Incident warning signs can be a useful addition to operations or team leader training. The objective of this training is for operations leaders to be able to recognize that identifying the incident warning signs is a proactive way to manage site and operational risks. They are also excellent leading indicators in a performance management process.

Specifically designed training exercises using the incident warning signs can help all levels of site leadership and supervision identify early warning of potential operating risk.

Training also provides leaders with a greater level of understanding in the importance of site visits and audits. Many of the incident warning signs are visible only when leaders are in the process units.

Some examples of incident warning signs training are:

- *Simulated exercises*: Take the participants through scenarios where incident warning signs are evident, but if they are not managed, the signs increase risk and can lead to an incident.
- *Use of photographs*: Ask the participants to identify the warning signs visible in the photograph. This is particularly helpful in engaging site leaders in recognizing that incident warning signs may exist at their site. These can be used in other training programs or safety moments to start a meeting as well.
- *Relate warning signs to the incident barrier model*: Emphasize that the incident warning signs are actually early warning that process safety barriers are starting to fail or have failed. Use an exercise asking participants to match the warning sign to the management system failure that it could indicate. These exercises are particularly helpful for site leaders to understand process safety barriers

12.3.5 All employees

Everyone has responsibilities for process safety. All employees regardless of position should ask the following questions:

- Do I keep my eyes open for warning signs daily?
- Do I report all incidents and abnormal conditions immediately?
- Do I follow all established policies and procedures 100% of the time?
- Does my organization follow all established policies and procedures 100% of the time?
- Do I feel a personal responsibility to correct unsafe conditions and stop unsafe work?
- Do I wear appropriate personnel protective equipment?
- Do I complete all prescribed training on time?
- Do I participate in regular emergency response drills?

12.4 SUMMARY

Managing risk from the perspective of process safety is everyone's responsibility. The warning signs in this book can help in predicting when your systems may be weak. These advance warnings provide an opportunity to improve and strengthen the systems that defend against catastrophic events. The action is up to you.

The following case study illustrates the importance of addressing the process safety elements discussed in the preceding chapters. It demonstrates how significant management system failures tend to overlap and reinforce one another.

12.5 CASE STUDY – OIL PLATFORM EXPLOSION AND FIRE IN THE NORTH SEA

On July 6, 1988, the Piper Alpha oil platform experienced a series of catastrophic explosions and fires. This platform, located in the North Sea approximately 110 miles from Aberdeen, Scotland, had 226 people on board at the time of the incident. Of that total, 165 perished along with two emergency response personnel who died during a rescue attempt. The platform was destroyed. Subsequent investigation was hindered by a lack of physical evidence. Based upon eyewitness accounts, it was concluded that, most likely, a release of light hydrocarbons (condensate made up of propane, butane, and pentane) occurred when a pump was restarted after maintenance. Unknown to the personnel starting the pump, a relief valve (RV) in the pump discharge had been removed for service and a plate had been loosely installed in its place on the piping flange (which was not readily visible from the pump). Upon restart of the pump, this flange leaked, producing a flammable vapor cloud, which subsequently found an ignition source.

The Piper Alpha platform was at the hub of a network of platforms interconnected by oil and gas pipelines. The initial explosion ruptured oil lines on Piper Alpha. The still-pressurized inter-platform pipelines fed the leaks. Managers on other platforms, aware of a problem on Piper Alpha (but not its severity), assumed that they would be instructed to shut down their operations, if needed. However, the explosion had interrupted communications from Piper Alpha and considerable intervals passed before these other platforms were isolated. A series of subsequent explosions occurred as the fires weakened natural gas riser manifolds at the base of the platform. The intensity of the fires prevented rescue efforts, either by helicopter or by ship. At the height of the event, natural gas was being burned on Piper Alpha at a rate equivalent to the entire United Kingdom natural gas consumption rate. Many of the platform crew retreated to the crew accommodation module, as they had been trained, to await evacuation. No organized attempt was made to retreat from the accommodation module, even though it became increasingly apparent that the conditions in the module were becoming untenable.

The subsequent investigation revealed the following:

- Two separate work permits had been issued for the condensate pump, one for the pump repair and one for testing the RV. The RV job had not been completed by the end of the shift and, rather than working overtime to complete it, it was decided to terminate the permit for that day and continue on the next. The craft supervisor suspended the permit and returned it to the control room without notifying operations staff of the job status.
- During shift turnover, the status of the pump work was addressed, but no mention was made of the RV work, and there was no mention of it in the control room or maintenance logs. Continuing problems with the adequacy of turnovers and log entries were a problem known to some. One staff member is quoted: "It was a surprise when you found out some things which were going on."
- The work permits for the pump and the RV did not reference each other, and it is likely that the permits had been filed in separate locations (one in the control room and one in the safety office). When the on-line condensate pump failed later in the shift, creating an imperative to start the spare to enable continued production, control room personnel were only aware of the pump repair work permit and proceeded to have the pump returned to service.
- The permit to work (PTW) system was often not implemented according to procedure.
- The diesel-powered fire pumps had been placed in manual control mode due to the presence of divers in the water around the platform. This practice was more conservative than company policies, and a 1983 fire protection audit report had recommended that this practice be discontinued. Placing the pumps in manual meant that personnel would have had to reach the pumps to start them after the explosion. However, conditions prevented this and, as a result, the Piper Alpha deluge system was unavailable.

- Had firewater been available, its effectiveness might have been limited. Distribution piping, including that in the platform module where the fires were most severe, was badly corroded and pluggage of sprinkler heads was a known problem dating back to 1984. Various fixes had been attempted and a project to replace the fire protection piping had been initiated, but work was lagging behind schedule. Tests in May 1988 revealed that approximately 50% of the sprinkler heads in the subject module were plugged.

- To put the previous two observations in perspective, the structural steel on Piper Alpha had no fireproofing, and it was known (at least to management) that "…structural integrity could be lost within ten to fifteen minutes if a fire was fed from a large, pressurized hydrocarbon inventory."

- The investigation revealed that emergency response training given to new platform personnel was cursory and not provided uniformly. Workers required training if they had not been on Piper Alpha in the last six months. However, training was often waived even if the interval was considerably longer, or if the individual reported that he had previously worked offshore elsewhere. A number of survivors reported that they had never been trained on the location of the life rafts or how to launch them.

- Evacuation drills were not conducted weekly as required. (One six-month period recorded only thirteen drills.) No full-scale shutdown drill had been conducted in the three years prior to the explosion.

- Platform managers had not been trained on their response to such an emergency on another platform. (Note: The various platforms were owned or operated by different companies.)

- Approximately one year before the explosion, company management had been cautioned in an engineering report that a large fire from escaping gas could pose serious concerns with respect to the safe evacuation of the platform. However, management discounted the likelihood of such an event, believing that existing protective systems were adequate. In fact, the gas risers upstream of the emergency isolation valves on Piper Alpha were not protected against fire exposure and, because of the diameter and length of the inter-platform gas lines, several days would be required to depressurize the pipelines in the event of a breach. It was the failure of these lines that ultimately destroyed Piper Alpha and prevented its evacuation.

- The Cullen report provided critical commentary on what was judged to be inadequate management oversight and follow-up on each of the issues described above.

Exercise: Can you identify warning signs that may have preceded this incident?

APPENDIX A – INCIDENT WARNING SIGN SELF-ASSESSMENT TOOL

 An AIChE Technology Alliance Center for Chemical Process Safety	INCIDENT WARNING SIGN SELF-ASSESSMENT TOOL
Facility Name:	

Instructions: Select the ranking number that best represents your perception that the warning sign is an issue.

5 – strongly agree / 4 – agree / 3 – neutral / 2 – disagree / 1 – strongly disagree

CATEGORY / WARNING SIGN	RANK
LEADERSHIP AND CULTURE	
Operating outside the safe operating envelope is accepted	
Job roles and responsibilities not well defined, confusing, or unclear	
Negative external complaints	
Signs of worker fatigue	

CATEGORY / WARNING SIGN	RANK
Widespread confusion between occupational safety and process safety	
Frequent organizational changes	
Conflict between production goals and safety goals	
Process safety budget reduced	
Strained communications between management and workers	
Overdue process safety action items	
Slow management response to process safety concerns	
A perception that management does not listen	
A lack of trust in field supervision	
Employee opinion surveys give negative feedback	
Leadership behavior implies that public reputation is more important than process safety	
Conflicting job priorities	
Everyone is too busy	
Frequent changes in priorities	
Conflict between workers and management concerning working conditions	
Leaders obviously value activity-based behavior over outcome-based behavior	

CATEGORY / WARNING SIGN	RANK
Inappropriate supervisory behavior	
Supervisors and leaders not formally prepared for management roles	
A poorly defined chain of command	
Workers not aware of or not committed to standards	
Favoritism exists in the organization	
A high absenteeism rate	
An employee turnover issue exists	
Varying shift team operating practices and protocols	
Frequent changes in ownership	
TRAINING AND COMPETENCY	
No training on possible catastrophic events and their characteristics	
Poor training on hazards of the process operation and the materials involved	
An ineffective or nonexistent formal training program	
Inadequate training on facility chemical processes	
No formal training on process safety systems	
No competency register to indicate the level of competency achieved by each worker	

CATEGORY / WARNING SIGN	RANK
Inadequate formal training on process-specific equipment operation or maintenance	
Frequent performance errors apparent	
Signs of chaos during process upsets or unusual events	
Workers unfamiliar with facility equipment or procedures	
Frequent process upsets	
Training sessions canceled or postponed	
Procedures performed with a check-the-box mentality	
Long-term workers have not attended recent training	
Training records are not current or are incomplete	
Poor training attendance is tolerated	
Training materials not suitable or instructors not competent	
Inappropriate use or overuse of computer-based training	
PROCESS SAFETY INFORMATION	
Piping and instrument diagrams do not reflect current field conditions	
Incomplete documentation about safety systems	
Inadequate documentation of chemical hazards	
Low precision and accuracy of process safety information documentation other than piping and instrument diagrams	

CATEGORY / WARNING SIGN	RANK
Material safety data sheets or equipment data sheets not current	
Process safety information not readily available	
Incomplete electrical/hazardous area classification drawings	
Poor equipment labeling or tagging	
Inconsistent drawing formats and protocols	
Problems with document control for process safety information	
No formal ownership established for process safety information	
No process alarm management system	
PROCEDURES	
Procedures do not address all equipment required	
Procedures do not maintain a safe operating envelope	
Operators appear unfamiliar with procedures or how to use them	
A significant number of events, resulting in auto-initiated trips and shutdowns	
No system to gauge whether procedures have been followed	
Facility access procedures not consistently applied or enforced	

CATEGORY / WARNING SIGN	RANK
Inadequate shift turnover communication	
Poor-quality shift logs	
Failure to follow company procedures is tolerated	
Chronic problems with the work permit system	
Inadequate or poor quality procedures	
No system for determining which activities need written procedures	
No established administrative procedure and style guide for writing and revising procedures	
ASSET INTEGRITY	
Operation continues when safeguards are known to be impaired	
Overdue equipment inspections	
Relief valve testing overdue	
No formal maintenance program	
A run-to-failure philosophy exists	
Maintenance deferred until next budget cycle	
Preventive maintenance activities reduced to save money	
Broken or defective equipment not tagged and still in service	
Multiple and repetitive mechanical failures	

CATEGORY / WARNING SIGN	RANK
Corrosion and equipment deterioration evident	
A high frequency of leaks	
Installed equipment and hardware do not meet good engineering practices	
Improper application of equipment and hardware allowed	
Facility firewater used to cool process equipment	
Alarm and instrument management not adequately addressed	
Bypassed alarms and safety systems	
Process is operating with out-of-service safety instrumented systems and no risk assessment or management of change	
Critical safety systems not functioning properly or not tested	
Nuisance alarms and trips	
Inadequate practices for establishing equipment criticality	
Working on equipment that is in service	
Temporary or substandard repairs are prevalent	
Inconsistent preventive maintenance implementation	
Equipment repair records not up to date	
Chronic problems with the maintenance planning system	

CATEGORY / WARNING SIGN	RANK
No formal process to manage equipment deficiencies	
Maintenance jobs not adequately closed out	
ANALYZING RISK AND MANAGING CHANGE	
Weak process hazard analysis practices	
Out-of-service emergency standby systems	
Poor process hazard analysis action item follow-up	
Management of change system used only for major changes	
Backlog of incomplete management of change items	
Excessive delay in closing management of change action items to completion	
Organizational changes not subjected to management of change	
Frequent changes or disruptions in operating plan	
Risk assessments conducted to support decisions already made	
A sense that <u>we always do it this way</u>	
Management unwilling to consider change	
Management of change item review and approval lack structure and rigor	
Failure to recognize operational deviations and initiate management of change	

CATEGORY / WARNING SIGN	RANK
Original facility design used for current modifications	
Temporary changes made permanent without management of change	
Operating creep exists	
Process hazard analysis revalidations are not performed or are inadequate	
Instruments bypassed without adequate management of change	
Little or no corporate guidance on acceptable risk ranking methods	
Risk registry is poorly prepared, nonexistent, or unavailable	
No baseline risk profile for a facility	
Security protocols not enforced consistently	
AUDITS	
Repeat findings occur in subsequent audits	
Audits often lack field verification	
Findings from previous audits are still open	
Audits are not reviewed with management	
Inspections or audits result in significant findings	
Regulatory fines and citations have been received	

CATEGORY / WARNING SIGN	RANK
Negative external complaints are common	
Audits seem focused on good news	
Audit reports are not communicated to all people affected	
Corporate process safety management guidance does not match a site's culture and resources	
LEARNING FROM EXPERIENCE	
Failure to learn from previous incidents	
Frequent leaks or spills	
Frequent process upsets or off-specification product	
High contractor incident rates	
Abnormal instrument readings not recorded or investigated	
Equipment failures widespread and frequent	
Incident trend reports reflect only injuries or significant incidents	
Minor incidents are not reported	
Failure to report near misses and substandard conditions	
Superficial incident investigations result in improper findings	
Incident reports downplay impact	

CATEGORY / WARNING SIGN	RANK
Environmental performance does not meet regulations or company targets	
Incident trends and patterns apparent but not well tracked or analyzed	
Frequent activation of safety systems	
PHYSICAL WARNING SIGNS	
Worker or community complaints of unusual odors	
Equipment or structures show physical damage	
Equipment vibration outside acceptable ranges	
Obvious leaks and spills	
Dust buildup on flat surfaces and in buildings	
Inconsistent or incorrect use of personal protective equipment	
Missing or defective safety equipment	
Uncontrolled traffic movement within the facility	
Open and uncontrolled sources of ignition	
Project trailers located close to process facilities	
Plugged sewers and drainage systems	
Poor housekeeping accepted by workers and management	

CATEGORY / WARNING SIGN	RANK
Permanent and temporary working platforms not protected or monitored	
Open electrical panels and conduits	
Condensation apparent on inner walls and ceilings of process buildings	
Loose bolts and unsecured equipment components	

APPENDIX B – COMPOSITE LIST OF CATASTROPHIC INCIDENT WARNING SIGNS

LEADERSHIP AND CULTURE

1. Operating outside the safe operating envelope is accepted
2. Job roles and responsibilities not well defined, confusing, or unclear
3. Negative external complaints
4. Signs of worker fatigue
5. Widespread confusion between occupational safety and process safety
6. Frequent organizational changes
7. Conflict between production goals and safety goals
8. Process safety budget reduced
9. Strained communications between management and workers
10. Overdue process safety action items
11. Slow management response to process safety concerns
12. A perception that management does not listen
13. A lack of trust in field supervision
14. Employee opinion surveys give negative feedback
15. Leadership behavior implies that public reputation is more important than process safety
16. Conflicting job priorities
17. Everyone is too busy
18. Frequent changes in priorities
19. Conflict between workers and management concerning working conditions
20. Leaders obviously value activity-based behavior over outcome-based behavior
21. Inappropriate supervisory behavior
22. Supervisors and leaders not formally prepared for management roles
23. A poorly defined chain of command
24. Workers not aware of or not committed to standards
25. Favoritism exists in the organization

26. A high absenteeism rate
27. An employee turnover issue exists
28. Varying shift team operating practices and protocols
29. Frequent changes in ownership

TRAINING AND COMPETENCY

30. No training on possible catastrophic events and their characteristics
31. Poor training on hazards of the process operation and the materials involved
32. An ineffective or nonexistent formal training program
33. Inadequate training on facility chemical processes
34. No formal training on process safety systems
35. No competency register to indicate the level of competency achieved by each worker
36. Inadequate formal training on process-specific equipment operation or maintenance
37. Frequent performance errors apparent
38. Signs of chaos during process upsets or unusual events
39. Workers unfamiliar with facility equipment or procedures
40. Frequent process upsets
41. Training sessions canceled or postponed
42. Procedures performed with a check-the-box mentality
43. Long-term workers have not attended recent training
44. Training records are not current or are incomplete
45. Poor training attendance is tolerated
46. Training materials not suitable or instructors not competent
47. Inappropriate use or overuse of computer-based training

PROCESS SAFETY INFORMATION

48. Piping and instrument diagrams do not reflect current field conditions
49. Incomplete documentation about safety systems
50. Inadequate documentation of chemical hazards
51. Low-precision-and-accuracy-of-process-safety-information documentation other than piping and instrument diagrams
52. Material safety data sheets or equipment data sheets not current
53. Process safety information not readily available
54. Incomplete electrical/hazardous area classification drawings
55. Poor equipment labeling or tagging
56. Inconsistent drawing formats and protocols
57. Problems with document control for process safety information
58. No formal ownership established for process safety information
59. No process alarm management system

PROCEDURES

60. Procedures do not address all equipment required

61. Procedures do not maintain a safe operating envelope
62. Operators appear unfamiliar with procedures or how to use them
63. A significant number of events, resulting in auto-initiated trips and shutdowns
64. No system to gauge whether procedures have been followed
65. Facility access procedures not consistently applied or enforced
66. Inadequate shift turnover communication
67. Poor-quality shift logs
68. Failure to follow company procedures is tolerated
69. Chronic problems with the work permit system
70. Inadequate or poor quality procedures
71. No system for determining which activities need written procedures
72. No established administrative procedure and style guide for writing and revising procedures

ASSET INTEGRITY

73. Operation continues when safeguards are known to be impaired
74. Overdue equipment inspections
75. Relief valve testing overdue
76. No formal maintenance program
77. A run-to-failure philosophy exists
78. Maintenance deferred until next budget cycle
79. Preventive maintenance activities reduced to save money
80. Broken or defective equipment not tagged and still in service
81. Multiple and repetitive mechanical failures
82. Corrosion and equipment deterioration evident
83. A high frequency of leaks
84. Installed equipment and hardware do not meet good engineering practices
85. Improper application of equipment and hardware allowed
86. Facility firewater used to cool process equipment
87. Alarm and instrument management not adequately addressed
88. Bypassed alarms and safety systems
89. Process is operating with out-of-service safety instrumented systems and no risk assessment or management of change
90. Critical safety systems not functioning properly or not tested
91. Nuisance alarms and trips
92. Inadequate practices for establishing equipment criticality
93. Working on equipment that is in service
94. Temporary or substandard repairs are prevalent
95. Inconsistent preventive maintenance implementation
96. Equipment repair records not up to date
97. Chronic problems with the maintenance planning system
98. No formal process to manage equipment deficiencies
99. Maintenance jobs not adequately closed out

ANALYZING RISK AND MANAGING CHANGE

100. Weak process hazard analysis practices
101. Out-of-service emergency standby systems
102. Poor process hazard analysis action item follow-up
103. Management of change system used only for major changes
104. Backlog of incomplete management of change items
105. Excessive delay in closing management of change action items to completion
106. Organizational changes not subjected to management of change
107. Frequent changes or disruptions in operating plan
108. Risk assessments conducted to support decisions already made
109. A sense that we always do it this way
110. Management unwilling to consider change
111. Management of change item review and approval lack structure and rigor
112. Failure to recognize operational deviations and initiate management of change
113. Original facility design used for current modifications
114. Temporary changes made permanent without management of change
115. Operating creep exists
116. Process hazard analysis revalidations are not performed or are inadequate
117. Instruments bypassed without adequate management of change
118. Little or no corporate guidance on acceptable risk ranking methods
119. Risk registry is poorly prepared, nonexistent, or unavailable
120. No baseline risk profile for a facility
121. Security protocols not enforced consistently

AUDITS

122. Repeat findings occur in subsequent audits
123. Audits often lack field verification
124. Findings from previous audits are still open
125. Audits are not reviewed with management
126. Inspections or audits result in significant findings
127. Regulatory fines and citations have been received
128. Negative external complaints are common
129. Audits seem focused on good news
130. Audit reports are not communicated to all employees affected
131. Corporate process safety management guidance does not match a site's culture and resources

LEARNING FROM EXPERIENCE

132. Failure to learn from previous incidents
133. Frequent leaks or spills
134. Frequent process upsets or off-specification product

135. High contractor incident rates
136. Abnormal instrument readings not recorded or investigated
137. Equipment failures widespread and frequent
138. Incident trend reports reflect only injuries or significant incidents
139. Minor incidents are not reported
140. Failure to report near misses and substandard conditions
141. Superficial incident investigations result in improper findings
142. Incident reports downplay impact
143. Environmental performance does not meet regulations or company targets
144. Incident trends and patterns apparent but not well tracked or analyzed
145. Frequent activation of safety systems

PHYSICAL WARNING SIGNS

146. Worker or community complaints of unusual odors
147. Equipment or structures show physical damage
148. Equipment vibration outside acceptable ranges
149. Obvious leaks and spills
150. Dust build up on flat surfaces and in buildings
151. Inconsistent or incorrect use of personal protective equipment
152. Missing or defective safety equipment
153. Uncontrolled traffic movement within the facility
154. Open and uncontrolled sources of ignition
155. Project trailers located close to process facilities
156. Plugged sewers and drainage systems
157. Poor housekeeping accepted by workers and management
158. Permanent and temporary working platforms not protected or monitored
159. Open electrical panels and conduits
160. Condensation apparent on inner walls and ceilings of process buildings
161. Loose bolts and unsecured equipment components

REFERENCES AND SELECTED REGULATIONS

Chapter 1 – Introduction

A Practical Approach to Hazard Identification for Operations and Maintenance Workers. New York, NY: Center for Chemical Process Safety of the American Institute of Chemical Engineers, 2010.

Guidelines for Engineering Design for Process Safety. New York, NY: Center for Chemical Process Safety of the American Institute of Chemical Engineers, 1993.

Guidelines for Implementing Process Safety Management Systems. New York, NY: Center for Chemical Process Safety of the American Institute of Chemical Engineers, 1994.

Guidelines for Integrating Process Safety Management, Environment, Safety, Health and Quality. New York, NY: Center for Chemical Process Safety of the American Institute of Chemical Engineers, 1996.

Guidelines for Process Safety Fundamentals for General Plant Operations. New York, NY: Center for Chemical Process Safety of the American Institute of Chemical Engineers, 1995.

Guidelines for Process Safety in Batch Reaction Systems, New York, NY: Center for Chemical Process Safety of the American Institute of Chemical Engineers, 1999.

Guidelines for Process Safety in Outsourced Manufacturing Operations. New York, NY: Center for Chemical Process Safety of the American Institute of Chemical Engineers, 2000.

Guidelines for Risk Based Process Safety. New York, NY: Center for Chemical Process Safety of the American Institute of Chemical Engineers, 2007.

Guidelines for Technical Management of Chemical Process Safety. New York, NY: Center for Chemical Process Safety of the American Institute of Chemical Engineers, 1989.

Plant Guidelines for Technical Management of Chemical Process Safety. New York, NY: Center for Chemical Process Safety of the American Institute of Chemical Engineers, 1991.

The Business Case for Process Safety Management. New York, NY: Center for Chemical Process Safety of the American Institute of Chemical Engineers, 2003.

Sutton, Ian. 5th Annual Symposium, Mary Kay O'Connor Process Safety Center. October 29-30, 2002. Warning Flags over Your Organization Or: How Lucky Are You Feeling Today? Copyright © Sutton & Associates, 2002.

Vaughan, Diane. *The Challenger Launch Decision: Risky Technology, Culture, and Deviance at NASA.* Chicago, IL: The University of Chicago Press, 1996.

Union Carbide Corporation. Bhopal Methyl Isocyanate Incident Investigation Team Report, March 20, 1985. Danbury, CT: Union Carbide Corp., 1985.

Chapter 2 – Incident Mechanics

Environmental News Service. Chinese Petrochemical Explosion Spills Toxics in Songhua River. November 25, 2005.

Guidelines for Investigating Chemical Process Incidents, 2nd edition. New York, NY: Center for Chemical Process Safety of the American Institute of Chemical Engineers, 2003.

Guidelines for Process Safety in Outsourced Manufacturing Operations. New York, NY: Center for Chemical Process Safety of the American Institute of Chemical Engineers, 2000.

Reason, James. *Human Error.* New York, NY: Cambridge University Press, 1990.

Walter, Robert. *Discovering Operational Discipline: Facilitator's Guide.* Amherst, MA: HRD Press, 2002.

Chapter 3 – Leadership and Culture

Deming, W. Edwards. *Out of the Crisis.* Cambridge, MA: MIT Center for Advanced Engineering Study, 1986.

Hopkins, Andrew. *Safety, Culture and Risk: The Organisational Causes of Disasters.* CCH Australia, 2005.

Klein, James A. and Bruce K. Vaughen. A revised program for operational discipline. *Process Safety Progress.* Volume 27, Issue 1, pages 58–65. March 2008.

Kouzes, James M. and Barry Z. Possner. *The Leadership Challenge,* 4th edition. San Francisco, CA: Jossey-Bass, 2008.

The Report of the BP U.S. Refineries Independent Safety Review Panel. January, 2007.

Vaughan, Diane. *The Challenger Launch Decision: Risky Technology, Culture, and Deviance at NASA.* Chicago, IL: The University of Chicago Press, 1996.

Walter, Robert. *Discovering Operational Discipline.* Amherst, MA: HRD Press, 2002.

Chapter 4 – Training and Competency

Anderson, L. W. and David R. Krathwohl, et al. (Eds.) *A Taxonomy for Learning, Teaching, and Assessing: A Revision of Bloom's Taxonomy of Educational Objectives.* Boston, MA: Allyn & Bacon, Pearson Education Group, 2001.

Bloom, Benjamin S. *Taxonomy of Educational Objectives: The Classification of Educational Goals*; pp. 201–207. Susan Fauer Company, Inc., 1956.

Guidelines for Preventing Human Error in Process Safety. New York, NY: Center for Chemical Process Safety of the American Institute of Chemical Engineers, 2004.

Hopkins, Andrew. *Lessons from Longford: The Esso Gas Plant Explosion.* CCH Australia, 2000.

Instructor Skills Workshop, AntiEntropics, Inc., 2009.

Chapter 5 – Process Safety Information

ANSI/API RP 505. Recommended Practice for Classification of Locations for Electrical Installations at Petroleum Facilities Classified as Class I, Zone 0, and Zone 2, 1st edition. Washington, DC: American Petroleum Institute, 1997.

API RP 500 (R2002) Recommended Practice for Classification of Locations for Electrical Installations at Petroleum Facilities Classified as Class I, Division I and Division 2, 2nd edition. Washington, DC: American Petroleum Institute, 1997.

Guidelines for Process Safety Documentation. New York, NY: Center for Chemical Process Safety of the American Institute of Chemical Engineers, 1995.

Health and Safety Executive Investigation Report, *The Fire at Hickson and Welch Limited.* London: HMSO, 1994.

Chapter 6 –Procedures

Guidelines for Safe Process Operations and Maintenance. New York, NY: Center for Chemical Process Safety of the American Institute of Chemical Engineers, 1995.

Guidelines for Writing Effective Operating and Maintenance Procedures. New York, NY: Center for Chemical Process Safety of the American Institute of Chemical Engineers, 1996.

Schlager, Neil. *When Technology Fails: Chernobyl Accident, Ukraine*. Farmington Hills, MI: Gale, 1994.

Walter, Robert. *Procedure Writing Techniques Workshop*. AntiEntropics, Inc., 1994, 2010.

Chapter 7 – Asset Integrity

Guidelines for Design Solutions for Process Equipment Failures. New York, NY: Center for Chemical Process Safety of the American Institute of Chemical Engineers, 1998.

Guidelines for Implementation of Safe and Reliable Instrumented Protective Systems. New York, NY: Center for Chemical Process Safety of the American Institute of Chemical Engineers, 2007.

Guidelines for Improving Plant Reliability Through Data Collection and Analysis. New York, NY: Center for Chemical Process Safety of the American Institute of Chemical Engineers, 1998.

Guidelines for Mechanical Integrity. New York, NY: Center for Chemical Process Safety of the American Institute of Chemical Engineers, 2006.

Guidelines for Process Equipment Reliability Data with Data Tables. New York, NY: Center for Chemical Process Safety of the American Institute of Chemical Engineers, 1989.

Guidelines for Safe Automation of Chemical Processes. New York, NY: Center for Chemical Process Safety of the American Institute of Chemical Engineers, 1993.

Inherently Safer Processes: A Life Cycle Approach, 2nd edition. New York, NY: Center for Chemical Process Safety of the American Institute of Chemical Engineers, 2008.

Investigation Report, Refinery Fire Incident, Tosco Avon Refiner. U.S. Chemical Hazard Investigation Board, March 2001

Chapter 8 – Analyzing Risk and Managing Change

Guidelines for Chemical Process Quantitative Risk Analysis. New York, NY: Center for Chemical Process Safety of the American Institute of Chemical Engineers, 1999.

Guidelines for Consequence Analysis of Chemical Releases. New York, NY: Center for Chemical Process Safety of the American Institute of Chemical Engineers, 1995.

Guidelines for Evaluating Process Plant Buildings for External Explosions and Fires. New York, NY: Center for Chemical Process Safety of the American Institute of Chemical Engineers, 1996.

Guidelines for Facility Siting and Layout. New York, NY: Center for Chemical Process Safety of the American Institute of Chemical Engineers, 2003.

Guidelines for Hazard Evaluation Procedures. New York, NY: Center for Chemical Process Safety of the American Institute of Chemical Engineers, 1995.

Guidelines for Management of Change for Process Safety. New York, NY: Center for Chemical Process Safety of the American Institute of Chemical Engineers, 2008.

Guidelines for Performing Effective Pre-Startup Safety Reviews. New York, NY: Center for Chemical Process Safety of the American Institute of Chemical Engineers, 2007.

Health and Safety Executive, *The Flixborough Disaster: Report of the Court of Inquiry.* London: HMSO.

Layer of Protection Analysis: Simplified Process Risk Assessment. New York, NY: Center for Chemical Process Safety of the American Institute of Chemical Engineers, 2001.

Chapter 9 – Audits

Guidelines for Auditing Process Safety Management Systems. New York, NY: Center for Chemical Process Safety of the American Institute of Chemical Engineers, 1992.

Marshall, V. C. *The Allied Colloids Fire and Its Immediate Lessons.* IChemE Loss Prevention Bulletin, April, 1994.

Chapter 10 – Learning from Experience

Atherton, John and Frederick Gil. *Incidents That Define Process Safety.* New York, NY: American Institute of Chemical Engineers John Wiley and Sons, Inc., Center for Chemical Process Safety, 2006.

Columbia Accident Investigation Board. *Report of Columbia Accident Investigation Board,* Volume I, 2003.

Kletz, Trevor. *Lessons from Disaster: How Organizations Have No Memory and Accidents Recur.* Houston, TX: Gulf Professional Publishing, 1993.

Guidelines for Performing Effective Pre-Startup Safety Reviews. New York, NY: Center for Chemical Process Safety of the American Institute of Chemical Engineers, 2006.

Guidelines for Evaluating the Characteristics of Vapor Cloud Explosions, Flash Fires & BLEVEs. New York, NY: Center for Chemical Process Safety of the American Institute of Chemical Engineers, 1994.

Guidelines for Fire Protection in the Chemical, Petrochemical, and Petroleum Industries. New York, NY: Center for Chemical Process Safety of the American Institute of Chemical Engineers, 2003.

Guidelines for Investigating Chemical Process Incidents, 2nd edition. New York, NY: Center for Chemical Process Safety of the American Institute of Chemical Engineers, 2003.

Guidelines for Post-release Mitigation Technology in the Chemical Process Industry. New York, NY: Center for Chemical Process Safety of the American Institute of Chemical Engineers, 1996.

Guidelines for Technical Planning for On-Site Emergencies. New York, NY: Center for Chemical Process Safety of the American Institute of Chemical Engineers, 1995.

Weick, Karl E. and Kathleen M. Sutcliffe. *Managing the Unexpected: Resilient Performance in an Age of Uncertainty*, 2nd edition. Hoboken, NJ: Jossey-Bass, John Wiley & Sons, Inc. 2007.

Chapter 11 – Physical Warning Signs

Guidelines for Design Solutions for Process Equipment Failures. New York, NY: Center for Chemical Process Safety of the American Institute of Chemical Engineers, 1998.

Guidelines for Implementation of Safe and Reliable Instrumented Protective Systems. New York, NY: Center for Chemical Process Safety of the American Institute of Chemical Engineers, 2007.

Guidelines for Improving Plant ReliabilityThrough Data Collection and Analysis. New York, NY: Center for Chemical Process Safety of the American Institute of Chemical Engineers, 1998.

Guidelines for Mechanical Integrity. New York, NY: Center for Chemical Process Safety of the American Institute of Chemical Engineers, 2006.

Investigation Report, Combustible Dust Fire and Explosions at CTA Acoustics Inc. U.S. Chemical and Hazard Investigation Board, February, 2005.

Chapter 12 – A Call to Action

Building Process Safety Culture: Tools to Enhance Process Safety Performance. New York, NY: Center for Chemical Process Safety of the American Institute of Chemical Engineers, 2005.

Cullen, The Honourable Lord. *The Public Inquiry into the Piper Alpha Disaster.*, London HMO, 1990.

Krause, Thomas *Catastrophic Events: Eight Questions Every Senior Leader Should Ask.* BST white paper, 2010.

Selected Regulations

- Occupational Safety and Health Administration (OSHA), *Process Safety Management of Highly Hazardous Chemicals,* 29 CFR Part 1910, Section 119 (Washington, DC, 1992)
- Environmental Protection Agency (EPA), *Accidental Release Prevention Requirement/Risk Management Programs,* Clean Air Act, Section 112 (r)(7) (Washington, DC, 1996)
- New Jersey Department of Environmental Protection, *Toxic Catastrophe Prevention Act (TCPA)*, N.J.S.A. 13:1K-19 et seq., 1986
- Pipeline and Hazardous Materials Safety Administration (PHMSA), *Gas Transmission Integrity Management Rule,* 49 CFR Part 192, Subpart O
- Pipeline and Hazardous Materials Safety Administration (PHMSA), *Liquid Pipeline Integrity Management in High Consequence Areas for Hazardous Liquid Operators.* 49 CFR Parts 195.450 and.195.452
- *Environmental Emergency Regulation,* (SOR/2003-307), Environment Canada
- *Control of Major Accident Hazards Involving Dangerous Substances*, European Directive Seveso II (96/82/EC)
- Korean OSHA PSM standard, *Industrial Safety and Health Act, Article 20, Preparation of Safety and Health Management Regulations*, Korean Ministry of Environment, Framework Plan on Hazardous Chemicals Management, 2001–2005
- Malaysia, Department of Occupational Safety and Health (DOSH) Ministry of Human Resources Malaysia, Section 16 of Act 514
- Mexican Integral Security and Environmental Management System (SIASPA), 1998
- *Control of Major Accident Hazards Regulations (COMAH),* United Kingdom Health & Safety Executive, 1999 and 2005

ACRONYMS AND ABBREVIATIONS

AIChE—American Institute of Chemical Engineers

BLEVE—boiling liquid expanding vapor explosion

CBT—computer-based training

CCPS—Center for Chemical Process Safety

CFR—*Code of Federal Regulations* (U.S.)

CM—corrective maintenance

CPI—chemical process industries

CSB—Chemical Safety Board (U.S.)

CUI—corrosion under insulation

DCS—distributed control system

DOT—Department of Transportation (U.S.)

EDMS—electronic document management system

EHS—environmental, health, and safety

EHSCD—environmental health and safety critical devices (EHSCDs)

EPA—Environmental Protection Agency (U.S.)

HAZOP—hazard and operability study

HEPA—high-efficiency particulate air (filter)

HSE—health and safety executive

HVAC—heating, ventilating, and air conditioning

ISO—International Organization for Standardization

KPI—key performance indicator

MIC— methyl isocyanate

MOC—management of change

MSDS—material safety data sheet(s)

NEC—National Electrical Code (U.S.)

OOS—out of service

OSHA—Occupational Safety and Health Administration (U.S.)

P&ID—piping (or process) and instrumentation diagram

PdM—predictive maintenance

PFD—process flow diagram

PHA—process hazard analysis

PM—preventive maintenance

PPE—personal protective equipment

PSI—process safety information

PSM—process safety management

PSSR—pre-startup safety review

PTW—permit to work

PVB—pressure vessel burst

QA—quality assurance

QC—quality control

RBPS— Risk Based Process Safety

RCM—reliability-centered maintenance

RMP—risk management program

RV—relief valve

SCBA—self-contained breathing apparatus

SIS—safety instrumentation systems

SME—subject matter expert

SOE—safe operating envelope

VCE—vapor cloud explosion

GLOSSARY

A

Abnormal Situation—disturbance in an industrial process with which the basic process control system of the process cannot cope. In the context of hazard evaluation procedures, synonymous with deviation.

Acceptable Risk—the average rate of loss that is considered tolerable for a given activity.

Accident—an incident that results in significant human loss.

Administrative Controls—procedural requirements for directing or checking engineered systems or human performance associated with plant operations.

Alarm Management—the set of processes and practices for determining, documenting, designing, monitoring, and maintaining alarm messages.

Audit—a systematic, independent review to verify conformance with prescribed standards of care using a well-defined review process to ensure consistency and to allow the auditor to reach defensible conclusions.

B

Baseline Risk Assessment—a process to characterize the current and potential threats to human health and the environment that may be posed by contaminants migrating to groundwater or surface water; releasing to air; leaching through soil; remaining in the soil and bio-accumulating in the food chain. The primary purpose of the baseline risk assessment is to provide risk managers with an understanding of the actual and potential risks to human health and the environment posed by the site and any uncertainties associated with the assessment. This information may be useful in determining whether a current or potential threat to human health or the environment warrants remedial action.

C

Catastrophic—a loss with major consequences and unacceptable lasting effects, usually involving significant harm to humans, substantial damage to the environment, and/or loss of community trust with possible loss of franchise to operate.

Configuration Management—the systematic application of management policies, procedures, and practices to assess and control changes to the hardware and software of a system and to maintain traceability of the configuration to the design basis throughout the system life. Configuration management is a specialized form of management of change.

Consequence(s)—the cumulative, undesirable result of an incident, usually measured in health and safety effects, environmental impacts, loss of property, and business interruption costs.

Controlled Document—documents covered under a revision control process to ensure that up-to-date documents are available and out-of-date documents are removed from circulation.

Core Value—a value that has been promoted to an ethical imperative, accompanied by a strong individual and group intolerance for poor performance or violations of standards for activities that the core value.

Corrective Maintenance—maintenance performed to repair a detected fault.

Critical Equipment—equipment, instrumentation, controls, or systems whose malfunction or failure would likely result in a catastrophic release of highly hazardous chemicals, or whose proper operation is required to mitigate the consequences of such release. (Examples are most safety systems, such as area LEL monitors, fire protection systems such as deluge or underground systems, and key operational equipment usually handling high pressures or large volumes.)

D

E

Effectiveness—the combination of process safety management performance and process safety management efficiency. An effective process safety management program produces the required work products of sufficient quality while consuming the minimum amount of resources.

Element Owner—the person charged with overall responsibility for overseeing a particular RBPS element. This role is normally assigned to someone who has management or technical oversight of the bulk of the work activities associated with the element, not necessarily someone who performs the work activities on a day-to-day basis.

Emergency Response Plan—a written plan which addresses actions to take in case of plant fire, explosion, or accidental chemical release.

Emergency Shutdown Device—a device that is designed to shut down the system to a safe condition on command from the emergency shutdown system.

Engineered Control—a specific hardware or software system designed to maintain a process within safe operating limits, to shut it down safely in the event of a process upset, or to reduce human exposure to the effects of an upset.

F

G

Good Engineering Practices—engineering, operating, or maintenance activities based on established codes, standards, published technical reports, or recommended practices.

H

HVAC—the heating, ventilation, and air-conditioning system of a building.

I

Incident Investigation—the management process by which underlying causes of undesirable events are uncovered and steps are taken to prevent similar occurrences.

Incident Investigation Management System—a written document that defines the roles, responsibilities, protocols, and specific activities to be carried out by personnel performing an incident investigation.

Incident Investigation Team—a group of qualified people who examine an incident in a manner that is timely, objective, systematic, and technically sound to determine that factual information pertaining to the event is documented, probable cause(s) are ascertained, and complete technical understanding of such an event is achieved.

Incident Warning Sign—an indicator of a subtle problem that could lead to an incident.

Intermediates—materials from a process that are not yet completely finished product. They may be a mixture or compound.

J

Job Task Analysis—the analysis phase of the instructional systems design (ISD) model consists of a job task analysis based on the equipment, operations, tools, and materials to be used as well as the knowledge and skills and attitudes required for each job position.

K

L

Layer of Protection Analysis (LOPA)—a process (method, system) of evaluating the effectiveness of independent protection layer(s) in reducing the likelihood or severity of an undesirable event.

M

Management of Change—a system to identify, review, and approve all modifications to equipment, procedures, raw materials, and processing conditions, other than replacement in kind, prior to implementation to help ensure that changes to processes are properly analyzed (for example, for potential adverse impacts), documented, and communicated to employees affected.

Mechanical Integrity (MI)—a management system for ensuring the ongoing durability and functionality of equipment.

N

Normal Operation—the phase of process operation between the startup phase and shutdown phase. Any process operations that can be performed during this period to support continued operation within safe upper and lower operating limits is a normal operations task.

Normalization of Deviance—a gradual erosion of standards of performance because of increased tolerance of nonconformance.

O

Operating Procedures—written step-by-step instructions and associated information (cautions, notes, warnings) for safely performing a task within operating limits.

Outsourced Manufacturing—providing manufacturing services for a fee by a contractor to a company issuing a contract for those services. Services can include reaction processes, formulation, blending, mixing or size reduction, separation, agglomeration, packaging, repackaging, and others, or a combination of the above.

P

Performance Measure—a metric used to monitor or evaluate the operation of a program activity or management system.

Personal Protective Equipment (PPE)—equipment designed to protect employees from serious workplace injuries or illnesses resulting from contact with chemical, radiological, physical, electrical, mechanical, or other workplace hazards. Besides face shields, safety glasses, hard hats, and safety shoes, PPE includes a variety of devices and garments, such as goggles, coveralls, gloves, vests, earplugs, and respirators.

Pre-startup Safety Review (PSSR)—a final examination, initiated by a trigger event, prior to the use or reuse of a new or changed aspect of a process. It is also the term for the OSHA PSM and EPA RMP element that defines a management system for ensuring that new or modified processes are ready for startup.

Preventive Maintenance—inspection or testing conducted on equipment to detect impending or minor failures and restoring the proper condition of the equipment.

Process Flow Diagram (PFD)—a diagram that shows the material flow from one piece of equipment to the other in a process. It usually provides information about the pressure, temperature, composition, and flow rate of the various streams, heat duties of exchangers, and similar information pertaining to understanding and conceptualizing the process.

Process Hazard Analysis (PHA)—an organized effort to identify and evaluate hazards associated with chemical processes and operations to enable their control. This review normally involves the use of qualitative techniques to identify and assess the significance of hazards. Conclusions and appropriate recommendations are developed. Quantitative methods can be used to help prioritize risk reduction.

Process Safety Information (PSI)—physical, chemical, and toxicological information related to the chemicals, process, and equipment. It is used to document the configuration of a process, its characteristics, its limitations, and as data for process hazard analyses.

Q

Quality Assurance (QA)—activities performed to ensure that equipment is designed appropriately and to ensure that the design intent is not compromised throughout the equipment's entire life cycle.

R

Replacement in Kind—replacement that satisfies the design specifications.

Risk—a measure of potential loss (for example, human injury, environmental impact, economic penalty) in terms of the magnitude of the loss and the likelihood that the loss will occur.

Risk Analysis—the development of a qualitative or quantitative estimate of risk based on engineering evaluation and mathematical techniques (quantitative only) for combining estimates of event consequences, frequencies, and detectability.

Risk Based Process Safety—the CCPS's process safety management system approach, which uses risk-based strategies and implementation tactics that are commensurate with the risk-based need for process safety activities, availability of resources, and existing process safety culture to design, correct, and improve process safety management activities.

Risk Factor—along with the probability that an event will occur (risk) are those factors of behavior, lifestyle, environment, or heredity associated with increasing or decreasing that probability.

S

Safety Instrumented System (SIS)—the instrumentation, controls, and interlocks provided for safe operation of a process.

Scale-up—the steps involved in transferring a manufacturing process or section of a process from laboratory scale to the level of commercial production.

T

Trigger Event—any change being made to an existing process, or any new facility being added to a process or facility, or any other activity that a facility designated as needing a pre-startup safety review. One example of a non-change-related trigger event is performing a PSSR before restart after an emergency shutdown.

U

V

Verification Activity—a test, field observation, or other activity used to ensure that personnel have acquired necessary skills and knowledge following training.

W

Worst-Case Scenario (WCS)—a release involving a hazardous material that would result in the worst (most severe) off-site consequences.

X

Y

Z

INDEX

D

E

K

L

M

N

O